D0901862

THE SCIENTIFIC COMMUNITY

Warren O. Hagstrom

SOUTHERN ILLINOIS UNIVERSITY PRESS

Carbondale and Edwardsville

FEFFER & SIMONS, INC.

London and Amsterdam

Library of Congress Cataloging in Publication Data
Hagstrom, Warren O
The scientific community.
(Arcturus books, AB130)
Reprint of the ed. published by Basic Books, New York.
1. Research. 2. Scientists. I. Title.
Q180.A1H3 1975 507'.2 74-18379
ISBN 0-8093-0720-0

Reprinted by arrangement with Basic Books, Inc.
Arcturus Books Edition February 1975
This edition printed by offset lithography
in the United States of America

TO MY MOTHER

Borghild Ingebjørg Aune Hagstrom

ACKNOWLEDGMENTS

William Kornhauser, Hanan C. Selvin, and Lewis Feuer advised me at the inception of this study and made numerous suggestions in the course of it. They contributed many ideas and prevented me from committing theoretical, factual, and stylistic errors. Arlene K. Daniels read the entire manuscript closely, and it is wiser and clearer for her having done so. I have learned much of what I know of the history of science from Thomas S. Kuhn's writings and lectures; I only wish I could have followed the implications of some of his leading ideas more than I have. My intellectual debts to Robert K. Merton are great; he has clear priority for many of the propositions advanced here—a priority inadequately indicated by the numerous references to his writings.

Portions of this research were completed when I was a fellow of the Social Science Research Council, and its assistance is gratefully acknowledged.

To my wife I owe something of my understanding of the nature of gifts.

WARREN O. HAGSTROM

Madison, Wisconsin
January 1965

CONTENTS

TABLES AND FIGURE

It seems that there is something in the nature of cultural achievements in our society that ensures that their primary reward is unstable subjective success—i.e., fame—and that other supplementary guarantees of objective success associated with it are regarded as merely accessory. . . . Work is primarily motivated by the *desire for recognition.*

—Karl Mannheim

Looked at more closely, what sort of man is the scholar? In the first place, he is neither authoritative nor self-sufficient. He is industrious, submerges himself patiently in the rank and file, and is moderate in his ability and his demands. He has the instinct for what is necessary to his kind: for example, that amount of independence and green pasture without which there is no peace for work; those claims to honor and recognition . . . , the sunshine which radiates from a good reputation; and that constant affirmation of his worth and usefulness with which the inner distrust, at the bottom of the hearts of all unfree men and herd animals, must again and again be overcome

—Friedrich Nietzsche

THE SCIENTIFIC COMMUNITY

INTRODUCTION AND METHODOLOGICAL NOTE

This work is concerned with the influence of scientific colleagues on the conduct of one another's research. With few exceptions, the discussion is limited to basic research in experimental sciences with well-established theories. In this type of research, the scientific community is relatively autonomous, and the group of colleagues is the most important source of social influence on research. Colleagues influence decisions to select problems and techniques, to publish results, and to accept theories.

Decisions such as these involve the central goals and values of the scientific community. Other aspects of scientific life are discussed only with reference to them. I am not concerned here with such topics as the personalities of scientists, their nonscientific backgrounds, their politics, or the consequences of their work for nonscientists, except insofar as they affect purely scientific activities. Such topics as the religious beliefs of scientists or their political activities are interesting, but they are interesting primarily from the point of view of religion and politics.

Scientists influence their colleagues in many ways, but I am concerned primarily with those influences on behavior that are normatively important to scientists. In other words, I am concerned with the operation of social control within the scientific community, with the problem of discovering the social influences that produce conformity to scientific norms and values.

In the first chapter, a theory of social control in science is presented. At the heart of this theory is the proposition that scientists are influenced by their desire to obtain recognition from colleagues. In Chapter II, the causes and consequences of competition for recognition are discussed. When scientists collaborate on the same problem, the question of apportioning rewards arises; this is discussed in Chapter III, along with the question of authority in collective efforts. The nature of the subcommunities of colleagues within which recognition is awarded is discussed in chapters IV and V, in the context of a theory of change in such communities.

A theory of social organization implies a theory of social disorganization. Three types of disorganization are described. The disorganizing effects of bureaucracy are analyzed in Chapter III, "anomy" in science, in Chapter IV, and substantive disputes, in Chapter VI.

The evidence for the arguments presented here comes from the following sources: previous work by sociologists and historians of science; publications by scientists themselves, whether autobiographical, concerned with organizational problems, or strictly scientific; my own field research; and my secondary analyses of data collected by other sociologists.

Sociology and the History of Science

Although sociologists have been interested in the organization of science from the time of Auguste Comte, most of the available research has been published since 1945.[1] Much of this research deals with the recruitment of scientists and with the organization of applied research and is only indirectly relevant to this study. More relevant are studies of basic scientists based on historical materials[2] and recent surveys of scientists in universities and other organizations in which basic research is conducted. I have often referred to the work of historians of science, although most of their published work is not about informal social organization; historians of science have been far more interested in the history of ideas than in the social organization of science.[3]

Publications by Scientists

From the time of Charles Babbage,[4] and even before him, scientists themselves have written perceptive accounts of the organization of science. Such accounts, however, seldom seek or produce sociological generalizations. Scientists' autobiographies are often unsatisfactory for sociological purposes. When scientists write such things, they carry with them the scientific norms of objectivity and detachment; since their social relations are not expected to influence their scientific judgments, they fail to report much about them. These statements by Charles Darwin and Albert Einstein are typical:

> My chief enjoyment and sole employment throughout life has been scientific work; and the excitement from such work makes me for the time forget, or drives away, my daily discomfort. I have therefore nothing to record during the rest of my life [from the time he moved to Down in 1842], except the publication of my several books.[5]

> [The] essential in the being of a man of my type lies precisely in *what* he thinks and *how* he thinks, not in what he does or suffers. Consequently, [this autobiographical essay] can limit itself in the main to the communicating of thoughts which have played a considerable role in my endeavors.[6]

Despite these disclaimers, the accounts of such scientists as Darwin and Einstein contain much about their social relations that is useful. In addition, research reports by scientists sometimes contain valuable information about the conduct of disputes, the award of recognition, and other matters of sociological interest.

Field Research for This Study

The major source of data consists of ninety focused interviews I conducted with a sample of scientists and science students. These interviews are used in two ways. Sometimes scientists are *quoted* as more or less expert *informants* about the scientific community. At other times they are *counted* as *respondents* in a social survey. In order to evaluate the uses I have made of the interviews in either of these ways, it is necessary to consider how the sample was selected.

The Sample

Since this is a study of the informal organization of basic science, the sample represents informal organization in its "purest" form; I wanted to avoid situations in which formal organization and external influences might conceal the effects of informal relations. Thus, the sample of seventy-nine professional scientists consists largely of members of university faculties who specialize in the "exact" sciences. All but three are members of university faculties; the others include two physicists working in an applied research institute associated with a university and a physicist on a postdoctoral fellowship. The sample was distributed among five universities, four of them among the top twenty-five in the United States.

Research in the exact sciences is least likely to be affected by "external" influences—the practical needs of consumers or the social and political convictions of scientists.[7] The exact sciences are those in which theoretical arguments are characterized by mathematical form and logical rigor and in which empirical confirmation of theories can be obtained from tightly controlled experiments or their equivalents. They include such fields as mathematics, physics, chemistry, astronomy, and experimental biology. With a small sample, it was originally decided to focus attention on a few disciplines, some of which would be well established and others "new." For example, mathematics and physics are well established, whereas statistics and molecular biology are relatively new. It was also decided to compare a discipline in which the exigencies of research require some formal organization with one in which there were no such exigencies; physics is an example of the former, mathematics of the latter. Thus, the range of disciplines represented in this study run from the formal sciences, through physics (and some related specialties in other disciplines), to molecular biology.

The formal sciences include mathematics, logic, and statistics and are distinguished from the empirical sciences in that their theories are not subject to empirical proof. Many philosophers would restrict the use of the term "science" to the empirical sciences. Although the philosophical difference between mathematics and the empirical sciences is of great importance, their social organization will be seen to be very similar.

The sample includes thirty-three formal scientists, most of whom are employed in departments of mathematics and the rest in philosophy and statistics departments.

The sample includes twenty-seven physical scientists. The original intention was to restrict the study of physical scientists to physicists. However, after interviewing several solid-state physicists, it was decided to include physical chemists who worked on closely related topics.

The biological scientists interviewed were mostly molecular biologists or men working in closely related areas. Molecular biologists study life processes at the level of the functioning molecule, by means of modern chemical and physical techniques. The field can be distinguished from classical biochemistry, on the one hand, and physiology, on the other. Classical biochemists analyze such substances as vitamins, which play a vital role in life processes, but they do not study the functioning of these substances in living organisms. Physiologists are concerned with how organisms function and are interested in the role of such substances as hormones in this process, but they do not subject these substances to detailed chemical analysis. These are somewhat oversimplified distinctions, since the research of biochemists, molecular biologists, and physiologists does overlap. For purposes of simplification and, in some cases, to conceal the identities of respondents, I have used the expression "molecular biology" for research even when the scientist involved might prefer calling it biochemistry or cell physiology.

There are few university departments of "molecular biology," although the number is growing. Specialists in this field are usually employed in departments of chemistry, biochemistry, bacteriology, microbiology, general biology, botany, zoology, virology, and perhaps physiology.

An early version of the interview guide was tried out on a sample of graduate students that included four physicists, five mathematicians, and two logicians. Students are sometimes quoted, but they were not counted in computing measures of correlation or in preparing tables.

The sample of professional scientists is biased to include a disproportionate number of eminent men. When convenient, I interviewed men who had been involved in scientific revolutions or had made great discoveries. (See the appendix to Chapter I for details.) I also attempted to include a number of formal leaders in the sample: eleven department chairmen; seven higher university officials; and fifteen scientists, who have been journal editors, officials of scientific societies, or active on the consulting bodies of grant-giving agencies.

This sample was designed for exploratory purposes and is not random. The results of one interview often influenced the selection of scientists for succeeding interviews. For example, after "algebraic topology" had been identified by an informant as a very popular field in mathematics, an attempt was made to interview a number of mathematicians who were either algebraic topologists or worked in closely related areas, and, after I got an idea of the kind of research in which solid-state physicists are engaged, I attempted to interview physical chemists who seemed to be doing research on similar topics. However, an attempt was made to get a range of specialists in each of the disciplines covered. *American Men of Science*, college catalogues, and abstracts of publications were used as aids in sampling.

Table 1 presents the sample by scientific field and type of position. It should be remembered that a person's scientific field does not always match the university department in which he is employed.

The Interviews

The interviews were unstructured, and the interview guide was revised in the course of the study and often tailored to fit the particular respondent. In most cases, the following sequence of topics was used: (1) the current research of the respondent; (2) productivity; (3) collaborative practices; (4) communication within departments and beyond them; (5) experiences and concern with anticipation; (6) goal conflicts within the discipline; and (7) substantive controversies within the field. Sometimes

TABLE 1
FIELD AND STATUS OF THE SAMPLE

SCIENTIFIC FIELD	UNIVERSITY FACULTY	OTHER PROFESSIONALS*	GRADUATE STUDENTS	TECHNICIANS
Mathematics	19		5	
Statistics	7			
Logic	7		2	
TOTAL FORMAL SCIENCES	33		7	
Theoretical physics	10	3		
Experimental physics	12		4	1
Physical chemistry	5			
TOTAL PHYSICAL SCIENCES	27	3	4	1
Molecular biology	12			1
Organic chemistry	1			
Physiology	1			
Botany (nonexperimental)	1			
Metallurgy	1			
TOTAL	76	3	11	2

* Ph.D.'s engaged in research.

questions were dropped when they proved unfruitful; at other times more questions about the same topic were asked in later interviews than in earlier ones. For example, after I became informed about specific substantive disputes, I was able to ask more detailed questions about the topic of controversies.

With a few very eminent men, the usual sequence of topics was almost entirely abandoned, and the interviews concerned almost entirely the revolutions they had led, the discoveries they had made, or the controversies in which they had been engaged. Despite this, and despite the fact that the form of questions was often changed, a number of questions in essentially the same form were asked of almost the entire sample. These form the basis of some of the tables presented in the following chapters.

The interviews ranged in duration from about forty-five minutes to two hours; most of them lasted between fifty and seventy minutes. Respondents were assured that their identities and, insofar as possible, the identities of their institutions would be kept confidential. To assure anonymity, names are always omitted when interviews are quoted, and sometimes scientific specialties are concealed. On the other hand, there is a good deal of specialized technical terminology in the quoted material. Those familiar with science may find such details valuable; readers who disagree with the facts presented here must pay attention to such details. But, since this is not a popularization of science, technical terminology is almost never explained in the text. The reader who is unfamiliar with science can best consider such terminology a way of "labeling" the scientist who uses it. Sociologists are interested in the social difference between algebraic topology and point-set topology, not the mathematical differences; it is enough for us to know such things as that contemporary mathematicians consider the former a fad while they view the latter as a sterile topic. Sometimes, but not usually, detailed knowledge of the subject matter is necessary in making sociological judgments. The sociology of science does not differ in this respect from other branches of sociology. Knowing the details of the differences between deep-sea and lagoon fishing helps to explain different types of behavior in Pacific islanders, but usually we can get by with rudimentary information about such things. The sociologist need not be a mathematician to understand the social organization of mathematics any more than he need be a shark fisherman to understand the social organization of Pacific islanders.

Secondary Analyses

In 1959, Bernard Berelson directed a study of graduate education in the United States.[8] This study involved questionnaires mailed to samples of the graduate faculties in United States universities (excluding schools of law and medicine) and to recent recipients of the Ph.D. Through the kindness of Professor Berelson, I was given access to his data, and some tables prepared from them are presented in the following chapters. The sample of the graduate faculty includes a total of 1,820 persons, 677 of whom teach in the physical or biological sciences, and the sample of recent Ph.D.'s totals 2,331, of whom 914 received their doctorates in the physical or biological sciences.[9]

Robert Wenkert and Roderic Fredrickson, in 1962, conducted an extensive survey of the faculty of the University of California, Berkeley, primarily to determine the use and potential demand for computer services at the University.[10] Their sample of 890 included about half of the physical and biological scientists at the University and more than half of the scientists actively engaged in research. Many of the questions they asked were relevant to the present study, and they, too, were kind enough to allow me access to their data.

I assisted Professor Kornhauser in the preparation of his book, *Scientists in Industry: Conflict and Accommodation,*[11] and in a few instances I have used interview material collected for that study. Kornhauser's book is often cited to contrast the behavior of industrial scientists with that of university scientists.

In an exploratory study we do not usually know what questions to ask which persons. From such a study we can develop an idea of the social structure in which activities occur and of typical social processes that go on in the structure. Further research is always necessary to confirm such conclusions and to elaborate the details of social structure and social processes in the population studied.

NOTES

1. Some of the best of this research has been collected in Bernard Barber and Walter Hirsch, eds., *The Sociology of Science* (New York: The Free Press of Glencoe, 1962).
2. The work of Robert K. Merton has been especially helpful for this study and will be referred to frequently.
3. There is a growing number of exceptions to this assertion, such as A. Hunter Dupree's excellent history of the formal organization of United States governmental research, *Science in the Federal Government* (Cambridge, Mass.: Harvard University Press, 1957).
4. Charles Babbage, *Reflections on the Decline of Science in England and Some of Its Causes* (London: B. Fellowes, 1830).
5. Charles Darwin, *The Autobiography of Charles Darwin, 1809–1882,* Nora Barlow, ed. (London: Collins, 1958), pp. 115 f.
6. Albert Einstein, in Paul A. Schilpp, ed., *Albert Einstein: Philosopher-Scientist* (Evanston, Ill.: The Library of Living Philosophers, 1949), p. 33.
7. The sociologists of knowledge have usually excluded pure science from the sphere of activities that might be socially determined. See, for example, Karl Mannheim, *Ideology and Utopia* (New York: Harcourt, Brace, 1936), p. 169. I was originally motivated by a desire to test this assertion.
8. Bernard Berelson, *Graduate Education in the United States* (New York: McGraw-Hill, 1960).
9. For details see Berelson, *op. cit.*, pp. 275–281.
10. This is reported in Robert Wenkert, in consultation with Roderic Fredrickson, *Use of a University Computer Center* (Berkeley: University of California Survey Research Center, 1962).
11. William Kornhauser, *Scientists in Industry: Conflict and Accommodation* (Berkeley: University of California Press, 1962).

I

SOCIAL CONTROL IN SCIENCE

The study of social control in science involves the search for characteristic types of behavior that produce conformity to or deviance from scientific norms and values. Many scientists would assert that the study of social control is of little importance because there is no problem of deviation in science—no significant tendencies by scientists to deviate or to induce conformity in others. Let us consider how such a position may be developed before attempting to discover the sources of social control in science; it is unprofitable to study inherently trivial forms of behavior.

The Limits of Socialization

The socialization of scientists tends to produce persons who are so strongly committed to the central values of science that they unthinkingly accept them. Research as an activity comes to be "natural" for them: they find it self-evident that persons should be excited by discoveries, intensely interested in the detailed working of nature, and committed to the elaboration of theories that are of no use whatever in daily life. They develop a vocabulary of motives that makes curiosity about nature and an interest in understanding it an intrinsically important component of the human personality.[1]

These commitments are the outcome of a prolonged training process, lasting well into adult life, in which the student is effectively isolated from competing vocational and intellectual interests and in which he is extremely dependent on his teachers. Not only does the teacher control the fate of his student, determining whether he will be permitted to enter a scientific profession and, if so, at what kind of institution, but also the self-conception of the student is dependent on the response of the teacher: the teacher's evaluation tends to be taken by the student as an indication of what he "is." The peer group of students (at the graduate level) also reinforces commitment to scientific values.[2] Although science students are not isolated from other students in any formal way during the course of their training, the pressure of work and study and simple propinquity tend to produce informal groups composed of those in the same fields. In this respect, of course, there are large differences among scientific fields; laboratory sciences are perhaps more likely to generate social isolation. In any case, isolation and the acquisition of visible status symbols are not as important to the scientist as they are to the physician, for example. Medical education, with its emphasis on school-class solidarity, fraternal living, and distinctive garb and titles, needs to create a visible identity because the student physician confronts nonphysicians as an essential part of his task and must be able to have his expertise accepted at face value. The basic scientist does not face these challenges to his self-identity.[3] Even so, his evaluation of his own competence and of the importance of his work depends on the social validation of his teachers and his peers.

The content of the texts, lectures, and laboratory work presented in the course of higher education in the sciences integrates the general norms and values of science with a specific set of beliefs and techniques. Thomas S. Kuhn has aptly described this kind of education:

> Typically, the undergraduate and graduate student of chemistry, physics, astronomy, geology, or biology acquires the substance of his field from books written especially for students. Until he is ready, or very nearly ready, to commence work on his own dissertation, he is neither asked to attempt trial research projects nor exposed to the immediate products of research done by others—to, that is, the professional communications that scientists write for each other. There are no collections of "Readings" in the natural sciences. Nor is the science student encouraged to read the historical classics of his field—works in which he might discover other ways of regarding the problems discussed in his text, but in which he would also meet problems, concepts and standards of solution that his future profession has long since discarded and replaced.
>
> In contrast, the various texts that the student does encounter display different subject matters, rather than, as in many of the social sciences, exemplifying different approaches to a single problem field. Even books that compete for adoption in a single course differ mainly in level and in pedagogic detail, not in substance or conceptual structure. Last, but most important of all, is the characteristic technique of textbook presentation. Except in their occasional introductions, science textbooks do not describe the sorts of problems that the professional may be asked to solve and the variety of techniques available for their solution. Rather these books exhibit concrete problem-solutions that the profession has come to accept as

> paradigms, and they then ask the student, either with a pencil and paper or in the laboratory, to solve for himself problems very closely related both in method and substance to those which the text or accompanying lecture has led him through. Nothing could be better calculated to produce "mental sets" or "*Einstellungen.*" Only in their most elementary courses do other academic fields offer even a partial parallel.[4]

Deviation from vague norms is more likely than deviation from norms specified for a concrete set of practices. It follows that physical scientists are less likely to deviate from the norms of science and scholarship than are social scientists or humanists.

As the scientist begins his professional life, his tasks typically reinforce the beliefs he has acquired as a student. In normal scientific research[5] "the characteristic problems are almost always repetitions, with minor modifications, of problems that have been undertaken and partially resolved before."[6] What the scientist has learned usually "works," and his technical success, regardless of any social confirmation of it, reinforces his commitments.

The effects of scientific socialization are reinforced by a highly selective system of recruitment. Of the fraction of the population who enter college, only fractions of those interested are permitted to graduate in the exact sciences and enroll in graduate school. Attrition in graduate school tends to be high, and only the more competent and highly motivated students obtain the doctorate.[7] Among those who do obtain doctorates in science, only a fraction are permitted to enter careers in basic research; the rest become teachers, administrators, and applied scientists. Basic scientists, then, are a highly selected and highly socialized elite group.

The entire socialization and selection process tends to produce scientists who are "self-starting" and "self-controlling." A common view of the organization of science, held implicitly or explicitly by most scientists, is that these individual characteristics are sufficient to account for conformity to scientific values and norms. It is often asserted that the scientist does what he wishes to do, attempts to solve problems that are intrinsically interesting and important, and is guided by aesthetic considerations. His social relations with others either interfere with this or are happy, but secondary, consequences of it.

This highly individualistic view is incomplete. It leads to no propositions about the actual scientific community as we know it, except perhaps that the socialization of recruits plays an unusually important part in the community's activities, and the importance of socialization is also consistent with other theoretical approaches. Some facts, which will be presented more fully in succeeding chapters, are inconsistent with the view of the scientist assumed by an individualistic theory. For instance, scientists seek to publish their accomplishments and are greatly disturbed if proper recognition for them is not forthcoming. Moreover, scientists who experience prolonged isolation from their colleagues cease being productive. A more obvious objection to this individualistic approach is that scientists seldom consciously set to work on problems that they know others have solved. If scientists received their major gratifications from problem-solving alone, the mere fact that others have solved the problem should not deter them from solving it themselves.[8] (Although mountaineers receive egoistic gratifications from being the first to climb a peak, they receive similar gratifications from climbing peaks already climbed: it demonstrates their abilities. Similarly, the egoistic scientist can demonstrate his abilities by solving previously solved problems, yet he seldom chooses to do so, since he also desires social recognition for his discoveries.)

Not only is the extremely individualistic view directly controverted by obvious facts about the scientific community, but there is every theoretical reason to expect this to be so. First, the autonomy of the scientific community cannot be taken for granted; it must be maintained by internal social controls, among other things. Without them, scientists would tend to respond more readily to the goals and standards of nonscientists. Second, communities of autonomous specialists tend to be rigid; they incorporate new goals and standards only with difficulty, for the socialization that produces commitment to norms and values at the highest levels also produces commitment to more specific norms. The scientific training that produces committed scientists also tends to commit them to techniques and particular theories. Since change is intrinsic in any community incorporating scientific values, if science is to thrive, scientists must respond to discoveries by continually changing their goals, techniques, and theories. Third, commitments to norms tend to erode in the absence of reinforcement. Many of the procedures known collectively as the "scientific method" are important only because they make possible communication among scientists. In the absence of sanctions, deviance from such norms would be common.

We may conclude that the socialization of scientists must be supplemented by a dynamic system of social control, if the values and effectiveness of science are to be maintained. Negative arguments are unsatisfying; the best reason for studying social control in science is that it leads one to discover the characteristic tensions within the scientific community, and this endeavor makes meaningful many varieties of scientific behavior that are otherwise unseen or dismissed as idiosyncratic and the consequence of aberrant personalities.

The Social Recognition of Discovery

Manuscripts submitted to scientific periodicals are often called "contributions," and they are, in fact, gifts. Authors do not usually receive royalties or other payments, and their institutions may even be required to aid in the financial support of the periodical.[9] On the other hand, manuscripts for which the scientific authors do receive financial payments, such as textbooks and popularizations, are, if not despised, certainly held in much lower esteem than articles containing original research results.

Gift-giving by scientists is thus similar to one of the most common modes of allocating resources to science, for this often takes the form of gifts from wealthy individuals or organizations. This has been true from the time of Cosimo de Medici to today, the time of the Rockefeller and Ford foundations. The gift status of moneys spent by industrial firms and governments on research is ambiguous; usually money seems to be spent with specific goals in mind, but the vast sums spent on space programs, particle accelerators, radiotelescopes, and so forth often seem like a potlatch by the community of nations. Neil Smelser has suggested that the gift mode of exchange is typical not only of science but of all institutions concerned with the maintenance and transmission of common values, such as the family, religion, and communities.[10]

In general, the acceptance of a gift by an individual or a community implies a recognition of the status of the donor and the existence of certain kinds of reciprocal rights.[11] These reciprocal rights may be to a return gift of the same kind and value, as in many primitive economic systems, or to certain appropriate sentiments of gratitude and deference. In science, the acceptance by scientific journals of contributed manuscripts establishes the donor's status as a scientist—indeed, status as a scientist can be achieved *only* by such gift-giving—and it assures him of prestige within the scientific community. The remainder of this chapter concerns the nature and forms of this allocation of prestige.

The organization of science consists of an exchange of social recognition for information. But, as in all gift-giving, the expectation of return gifts (of recognition) cannot be publicly acknowledged as the motive for making the gift. A gift is supposed to be given, not in the expectation of a return, but as an expression of the sentiment of the donor toward the recipient. Thus, in the kula expeditions of the Melanesians:

> The ceremony of transfer is done with solemnity. The object given is disdained or suspect; it is not accepted until it is thrown on the ground. The donor affects an exaggerated modesty. Solemnly bearing his gift, accompanied by the blowing of a conch-shell, he apologizes for bringing only his leavings and throws the objects at his partner's feet. . . . Pains are taken to show one's freedom and autonomy as well as one's magnanimity, yet all the time one is actuated by the mechanisms of obligation which are resident in the gifts themselves.[12]

Gift-giving is capable of cynical manipulation; if this is publicly expressed, however, the exchange of gifts ceases, perhaps to be succeeded by contractual exchange. Consequently, scientists usually deny that they are strongly motivated by a desire for recognition, or that this desire influences their research decisions. A biochemist gave a typical reply when asked whether scientists compete for recognition:

> Most scientists have sincere interests in the advancement of science, more than in their own recognition. I don't think honor is the greatest ambition of professionals generally; rather, they want to solve a problem. Professionals are a relatively idealistic group. You might find extremists of all sorts, but by and large their real interest is in their work and in advancing knowledge. They are motivated by curiosity. These people think that rewards will take care of themselves; they are fatalistic in that respect. . . . There is, after all, a kind of achievement in just getting to be a professor [i.e., in attaining the status of a scholar-scientist in the larger community], which is likely to be satisfying.

Some of my informants allowed themselves to be pressed into admitting that recognition was a source of gratification. For example, a theoretical physicist responded, when told that another scientist was not disturbed at all when he was anticipated and thereby prevented from publishing:

> I think I would admit to not having such pure interests. I must admit to a desire for recognition. I suppose it doesn't make much difference whether one wants to glorify himself or have others glorify him.

It is only the exceptional scientist, however, who sees the desire for recognition as a prime motivating force for himself and his colleagues. A mathematician, for example, said:

> A field in mathematics may become popular if the more popular mathematicians, the big shots, become interested in it. Then it grows rapidly. Junior mathematicians want recognition from big shots and, consequently, work in areas prized by them.

This man was something of a social isolate, capable of taking a detached view of the system.[13]

Nevertheless, the public disavowal of the expectation of recognition in return for scientific contributions should no more be taken to mean that the expectation is absent than the magnanimous front of the kula trader can be taken to mean that he does not expect a return gift. In both instances, this is made clearest when the expected response is not forthcoming. In primitive societies, failure to present return gifts often means warfare.[14] In science, the failure to recognize discovery may give rise, if not to warfare, at least to strong antagonisms and, at times, to intense controversy. A historical summary and analysis of priority controversies has been given by Robert K. Merton,[15] who pointed out that the failure to recognize previous work threatens the system of incentives in science. The pattern is not infrequent today. Of my seventy-nine informants, at least nine admitted to having been involved in questions of disputed priority either as the culprit or the victim. (The question was not asked in all seventy-nine interviews.) For example, an eminent theoretical physicist testified as follows:

> [This priority dispute] happened through an unfortunate habit of mine of not publishing things, delaying a year or two, talking about them, but not publishing. Therefore it has not always been clear to me whether something was known as my work or not. Under those conditions it is rather easy to get in priority disputes. It is much more satisfactory if one simply publishes what one does as he goes along. But a feeling of perfectionism frequently interferes with that. You don't want to publish something that's wrong. So you do the work, think about it for a couple of years, talk about it to everyone, but don't publish it. By that time somebody else has thought about it, and of course it's impossible to tell whether there was any influence from your work or not. It's a situation that's not very good. . . . *Did this result in some hostility against you?*[16] Yes. One or two people—there haven't been many—were quite acrimonious. *They saw you about it?* Yes. . . . It was a completely personal matter, and unfortunate.[17]

Another man, a little-known experimental physicist, was the victim in such a situation. A departmental colleague told about it first, when he was asked about the consequences of failure to recognize work:

> It causes a fatalistic attitude. . . . A professor on our faculty concerned himself with the X effect long before it became popular in the early nineteen fifties. [He did his work in the late nineteen thirties.] His work was referred to *once* by a review from a large laboratory, and in the review his name was omitted—"a professor at Y university" was the way they expressed it. This was deplorable. No action was ever taken on that. The man was just disappointed.

The experimental physicist himself described the same sequence of events, omitting the specific failure to recognize his accomplishments. Something similar to this had happened earlier in his career, when a grant he had requested was rejected, and shortly afterward someone else had become famous for doing essentially what he had proposed to do. This scientist was the most secretive man interviewed and was relatively isolated from his colleagues. In this case the individual affronted took no action; he may have reacted with hostility, but this was not communicated to the offenders. In other cases, as noted, affronted parties will communicate directly with the offenders. This may lead to a public recognition of priority—a "Notice of Priority," as it is called in the mathematical periodicals—or to an expression of recognition in succeeding papers by the offender or his collaborators. Thus, a mathematical statistician said that he was once anticipated in "a serious matter":

> . . . I hastened to recognize the other. In 1934 . . . I overlooked a paper by a Russian, published twelve years before in an Italian journal. It was a

very messy paper. In 1950 somebody from Iowa wrote me and informed me of these results, so I wrote a note of "Recognition of Priority."

The desire to obtain recognition induces scientists to publish their results. "Writing up results" is considered to be one of the less pleasant aspects of research—it is not intrinsically gratifying in the way that other stages of a research project are. Some respondents were asked about the source of the greatest gratifications in research work. Generally the response was that most gratification came when the problem was essentially solved—when one became confident that an experiment would be successful, or when the outlines of a mathematical proof became clear—although details might remain to be cleaned up. For example, a theoretical physicist said: "[I get most pleasure] when the problem has been solved in principle but when some hard work remains to be done—when you have enough security to know you're not wasting your time but while there is some challenge left in the problem."

An experimental physicist said one receives most gratification: ". . . when you find the effect you're looking for—everything else is anticlimax. Also in seeing some new and unexpected effect—in seeing new phenomena before others do."

Research is in many ways a kind of game, a puzzle-solving operation in which the solution of the puzzle is its own reward.[18] "Everything else"—including the communication of results—"is anti-climax." Writing up results, "cleaning up loose ends," may be an irksome chore. A mathematical statistician said he was often pleased to discover that the result he had obtained was already in the literature:

> Being anticipated doesn't bother me. Maybe it does once in a while, when I have something really nice. But if someone else has published the result it means I don't have to write it out, and that is gratifying. The real pleasure in the work comes in working the problem out. Publishing is necessary for money and is nice in a way. I try to publish what I find amusing and what I hope others will find amusing as well. . . . Actually, I shouldn't give the impression that one publishes only in order to survive. People who are in really secure positions publish anyway. For example, [a very eminent man in the department] publishes about five papers annually, and he wouldn't have to publish at all any more.

The desire to obtain social recognition induces the scientist to conform to scientific norms by contributing his discoveries to the larger community. Thomas Sprat, writing near the dawn of modern science, perceived the importance of this: "If neither *Chance*, nor *Friendship*, nor *Treachery* of Servants, have brought such Things out; yet we see *Ostentation* alone to be every day powerful enough to do it. This Desire of Glory, and to be counted Authors, prevails on all. . . ."[19]

Not only does the desire for recognition induce the scientist to communicate his results; it also influences his selection of problems and methods. He will tend to select problems the solution of which will result in greater recognition, and he will tend to select methods that will make his work acceptable to his colleagues.

The range of acceptable methods varies. In mathematics, for example, the standards of rigor have changed steadily. E. T. Bell has pointed this out in noting that eighteenth-century mathematicians were lucky, by later standards, not to have made more mistakes than they did:

> How did the master analysts of the 18th century—the Bernoullies, Euler, Lagrange, Laplace—contrive to get consistently right results in by far the greater part of their work in both pure and applied mathematics? What these great mathematicians mistook for valid reasoning at the very beginning of the calculus is now universally regarded as unsound.[20]

In this field, and most others, the change of standards is one of progress; the later standards can be shown to be technically superior. However, the definition of appropriate standards is not a technical matter only. For example, one informant was relatively famous for a method he had devised for making a certain kind of biochemical analysis. This method depended on distinctive nutritional requirements of certain bacteria. He noted that his method, while clearly superior to its alternatives for some purposes, and while very widely used, tended to be neglected by chemists:

> [The method] fell into some disrepute because we're using organisms, and the chemists said "Huh. Nobody can do the quantitative work with an organism." . . . There are other methods which chemists adopt. Chemists are a peculiar breed. They feel it is slightly debasing to use organisms. They just couldn't do that. . . . Most chemists use [other methods] for psychological reasons if nothing else. But I would never agree that other methods are generally better, except for specific purposes.

That is, social recognition in biochemical work done by chemists induces them to use techniques defined as distinctively chemical.[21]

Similarly, in mathematics the style of a proof, its "elegance," is often considered as important to its merit as the truth of the theorem proved. While there are technical reasons for this, there are distinctively social ones as well. A mathematician described a mixed case:

> Let me tell you a story about a famous problem in topology called the Poincaré conjecture. Various proofs were talked about, maybe they were published, but errors were found in all of them. Then, three years ago a Japanese mathematician published a hundred-page proof of it. A hundred-page proof is very unusual. Nobody has gone through it yet and found an error—this would be very difficult—but nobody believes in it. The man didn't have any new methods, and people didn't think he could do it with what he had.

Conformity with methodological standards is necessary if social recognition is to be given for contributions.

Similarly, the goals of science as they are specified in particular disciplines at particular times cover a restricted range, and the process of the reward of social recognition tends to produce individual conformity to the differentiated goals. As John R. Baker has put it:

> The scientist is able to construct a sort of scale of scientific values and to decide that one thing or theory is relatively trivial and another relatively important, quite apart from any question of practical applications. There is, as Poincaré has well said, "une hierarchie des faits." Most scientists will agree that certain discoveries or propositions are more important because more widely significant than others, though around any particular level on the scale of values there may be disagreement.[22]

Erwin Schrödinger, the eminent theoretical physicist, has indicated that this has been true in his experience as well:

> . . . it might be said that scientists all the world over are fairly well agreed as to what further investigations in their respective branches of study would be appreciated or not. . . . The argument applies to the research workers all the world over, but only of *one* branch of science and of *one* epoch. These men practically form a unit. It is a relatively small community, though widely scattered, and modern methods of communication have knit it into one. The members read the same periodicals. They exchange ideas with one another. And the result is that there is a fairly definite agreement as to what opinions are sound on this point or that. . . . In this respect international science is like international sport. . . . Just as it would be useless for some athlete in the world of sport to puzzle his brain in order to initiate something new—for he would have little or no hope of being able to "put it over," as the saying is—so too it would, generally speaking, be a vain endeavor on the part of some scientist to strain his imaginative vision toward initiating a line of research hitherto not thought of.[23]

Sanctions to enforce conformity in this respect, as well as with regard to appropriate techniques, are of two general kinds. First, works that deviate too far from the norm will be refused publication in scientific journals. A mathematician interviewed had a paper rejected for this reason:

> A couple of us did some work that another person thought was merely trivial. And there was a little cat fighting going on here. . . . This was rejected by one journal, we felt for poor reasons; we rewrote it and sent it in again, and it was again rejected, we felt because of a sloppy refereeing job. I have been tempted and may yet write a complaining letter to the editor, whom I know, not asking that the paper be reconsidered but that the referee's work be evaluated, because the paper was not well refereed.[24]

Such an exercise of sanctions makes it impossible for the great mass of scientists to evaluate for themselves the importance and validity of the information presented. Delegating considerable power to a few authorities obviously infringes on the norms of independence in science.[25] For this reason, editors and referees tend to be tolerant, basing their decisions on estimates that others will find manuscripts interesting even if they themselves do not. This and the fact that there are many journals, some of them unrefereed,[26] means that the sanction is of little importance to most scientists most of the time. A more important sanction is the social recognition published work receives; this sanction is exercised by the community at large and applies to all published research.

Another type of sanction is not primarily important in science, although it is often alleged to be. This consists of extrinsic rewards, primarily position and money.[27] It is alleged that scientists publish, select problems, and select methods in order to maximize these rewards. University policies that base advancement and salary on quantity of publication sometimes seem to imply that this is true, that scientists' research contributions are not freely given gifts at all but are, instead, services in return for salary. While it is important for extrinsic rewards to be more or less consistent with recognition, the ideal seems to be that they should follow recognition, and this seems to be the general practice. In any case, an explanation of scientific behavior in terms of extrinsic rewards is weakened by the fact that many scientists in elite positions, whose extrinsic rewards will be unaffected by their behavior, continue to be highly productive and to conform to scientific goals and norms. Furthermore, scientists usually feel that it is degrading and improper to submit manuscripts for publication primarily to gain position without really caring if the work is read by others.

But why should gift-giving be important in science when it is essentially obsolete as a form of exchange in most other areas of modern life, especially the most distinctly "civilized" areas? Gift-giving, because it tends to create particularistic obligations, usually reduces the rationality of economic action. Rationality is maximized when "costs" of alternative courses of action can be assessed, and such costs are usually established in free-market exchanges or in the plans of central directing agencies. When participants are paid a money wage or salary for their efforts, and when this effectively controls their behavior, the system is more flexible than when controls derive from traditional or gift obligations.[28] Why, then, does this frequently inefficient and irrational form of control persist in science? To be sure, it also tends to persist in other professions. Professionals are expected to be motivated by a desire to serve others.[29] For example, physicians do receive a fee for service, yet they are expected to have a "sliding scale" and serve the indigent at reduced fees or for no fees at all. The larger community recognizes two types of public dependence on professions: professional services are regarded as essential, concerned with values that should be realized regardless of a client's ability to pay; and nonprofessionals are unable to evaluate professional services, which makes them vulnerable to exploitation by unqualified persons. The rationale for the norm of service is usually the former type of dependence. In science, for example, the fact that a community has no one willing and able to pay for an important item of useless knowledge is not supposed to interfere with its ability to acquire the knowledge. But the idea of the gift and the norm of service is also related to the dependence of the public that follows from its inability to evaluate services.

The rationality of professional services is not the same as the rationality of the market. In contractual exchanges, when services are rewarded on a direct financial or barter basis, the client abdicates, to a considerable degree, his *moral control* over the producer. In return, the client is freed from personal ties with the producer, and he is able to choose rationally between alternative sources of supply. In the professions, and especially in science, the abdication of moral control would disrupt the system. The producer of professional services must be strongly committed to higher values. He must be responsible for his products, and it is fitting that he not be alienated from them. The scientist, for example, must be concerned with maintaining and correcting existing theories in his field, and his work should be oriented to this end. The exchange of gifts for recognition tends to maintain such orientations. On the one hand, the recipient of the gifts finds it difficult to refuse them (they are "free"), and, on the other, the donor is held responsible for adhering to central norms and values. The maxim, *caveat emptor*, is inapplicable.[30] Furthermore, the donor is not alienated from his gift, but retains a lasting interest in it. It is, in a sense, his property.[31] One indication of this is the frequent practice of eponymy, the affixing of the name of the scientist to all or part of what he has found.[32]

Emphasis on gifts and services occurs frequently in social life, and we can get at the root of this generality by focusing on certain paradoxical elements implicit in the argument presented thus far. It has been argued that scientists are oriented to receiving recognition from colleagues and that this orientation influences their research decisions. Yet evidence that scientists themselves deny this has also been presented. There is a normative component to this denial, one that appears more clearly in analyzing scientific fashions. It is felt that, if a scientist's decisions are influenced by the probability of being recognized, he will tend to deviate from certain central scientific norms—he will fail to be original and critical. Thus, while it is true that scientists are motivated by a desire to obtain social recognition, and while it is true that only work on certain types of problems and with certain techniques will receive recognition at any particular time, it is also true that, if a scientist were to admit being influenced in his choices of problems and techniques by the probability of being recognized, he would be considered deviant. That is, if scientists conform to norms about problems and techniques as a result of this specific form of social control, they are thereby deviants.

This apparent paradox, that people deviate in the very act of conforming, is common whenever people are expected to be strongly committed to values. In general, *whenever strong commitments to values are expected, the rational calculation of punishments and rewards is regarded as an improper basis for making decisions.* Citizens who refrain from treason merely because it is against the law are, by that fact, of questionable loyalty; parents who refrain from incest merely because of fear of community reaction are, by that fact, unfit for parenthood; and scientists who select problems merely because they feel that in dealing with them they will receive greater recognition from colleagues are, by that fact, not "good" scientists. In all such cases the sanctions are of no obvious value: they evidently do not work for the deviants, and none of those who conform admit to being influenced by them. But this does not mean that the sanctions are of no importance; it does mean that more than overt conformity to norms is demanded, that inner conformity is regarded as equally, or more, important.

Thus, the gift exchange (or the norm of service), as opposed to barter or contractual exchange, is particularly well suited to social systems in which great reliance is placed on the ability of well-socialized persons to operate independently of formal controls. The prolonged and intensive socialization scientists experience is reinforced and complemented by their practice of the exchange of information for recognition. The socialization experience produces scientists who are strongly committed to the values of science and who need the esteem and approval of their peers. The reward of recognition for information reinforces this commitment but also makes it flexible. Recognition is given for kinds of contributions the scientific community finds valuable, and different kinds of contributions will be found valuable at different times.

The scientist's denial of recognition as an important incentive has other consequences related to those already mentioned. When peers exchange gifts, the denial of the expectation of reciprocity in kind implies an expectation of gratitude, a highly diffuse response.[33] It will be shown that this kind of gift exchange occurs among scientists, although the more important form of scientific contribution is directed to the larger scientific community.[34] In this case the denial of the pursuit of recognition serves to emphasize the universality of scientific standards: it is not a particular group of colleagues at a particular time that should be addressed, but all possible colleagues at all possible periods. These sentiments were expressed with his typical fervor by Johannes Kepler:

> I have robbed the golden vessels of the Egyptians to make out of them a tabernacle for my God, far from the frontiers of Egypt. If you forgive me, I shall rejoice. If you are angry, I shall bear it. Behold, I have cast the dice, and I am writing a book either for my contemporaries, or for posterity. It is all the same to me. It may wait a hundred years for a reader, since God has also waited six thousand years for a witness.[35]

While this orientation is consistent with the scientist's need for autonomy—being dependent on the favors of particular others is terrifying—it also contains a strong element of the tragic. Scientists learn to *expect* injustice, the inequitable allocation of rewards. Occasionally one of them makes this explicit. Max Weber addressed students on "Science as a Vocation" in the following way:

> I know of hardly any career on earth where chance plays such a role. . . . If the young scholar asks for my advice with regard to habilitation, the responsibility of encouraging him can hardly be borne . . . one must ask every . . . man: Do you in all conscience believe that you can stand seeing mediocrity after mediocrity, year after year, climb beyond you, without becoming embittered and without coming to grief? Naturally, one always receives the answer: "Of course, I live only for my calling." Yet, I have found that only a few men could endure this situation without coming to grief.[36]

More common than such an explicit statement is the myth of the hero who is recognized only after his death. This myth is important in science, as in art, because it strengthens universal standards against tendencies to become dependent on particular communities. Thus, mathematics has such heroes as Galois, who wrote the major part of his great mathematical opus the night before he was killed in a duel at the age of twenty-one; Abel, who died of tuberculosis as his greatness was coming to be recognized; and Cantor, who died mad, believing his ideas were spurned by others.[37] The stories of Copernicus, receiving his revolutionary book the day he died, and Mendel, rediscovered years after his death, are well known in the larger society, where they perhaps serve a function similar to that performed in science, albeit more general.[38]

The distinctive functions of the system by which gifts of information are exchanged for recognition can be seen within science. Textbook-writing and the preparation of popularizations are expected to be neither original in the scientific sense nor critical of existing theories. Since texts draw on what is already known, teachers who adopt them are usually competent to judge the validity of the material presented. Consequently, writing of this sort is regarded as a technical skill and not a highly responsible task, and it has little effect on a scientist's reputation.[39] A dean in a leading university said: "We classify text writing under a man's teaching, and it may help him [in appointment and promotion]. However, if he has written nothing at all but texts, they will have a null value or even negative value."[40]

The author of a leading mathematical text also thought there might be risk involved: "I don't think [writing the text] injured me, although my book came out in 19—, when I was an instructor, and I was told this might be foolish. It doesn't give you credit as a research mathematician. . . ."

Because authors of texts and popularizations need not be highly committed to the values of the scientific community, their activities have little bearing on their standing within it. As a result, incentives for such tasks must be of a more general nature. Royalties are such a generalized reward; unlike recognition, cash can be used outside the community of pure science. The exchange of information for recognition, on the other hand, binds donors and recipients in a community of values.

Thus, gift-giving in exchange for recognition is an appropriate method of social control in science, and it is apparently relatively effective.[41] Some of its limitations are considered in later sections. First, however, it is necessary to clarify the nature of scientific gifts and recognition. Information is not usually presented, ribbon-wrapped, to an individual recipient. Who, then, is the recipient? And how does he respond?

Elementary and Institutionalized Recognition

Because the source of recognition in science is the source of control, it is critical for the organization of science. Two forms of recognition can be distinguished. "Institutional recognition" is given in formal channels of communication in science, whereas interpersonal approval and esteem, or "elementary recognition," is given in direct communication. Elementary forms of recognition are inadequate for science as an institution, but they play an important role in mediating between the individual scientist and the larger scientific community.

INSTITUTIONALIZED RECOGNITION

Formal communication in the sciences is primarily carried on through articles appearing in scientific journals. Books are also important, but not as important as they once were, or as they are now in the social sciences. On some occasions, papers read at public meetings serve as a substitute for an article, but when they do not merely supplement articles they are usually a less "serious" form of communication. Such public addresses, like books, were more important in the past.

The information conveyed in an article is highly specific; usually one problem or a few related problems are discussed. Recognition of other scientists for their contributions is conveyed in the same channel of communication. It usually takes the form of either footnote citations to specific articles or a section on "acknowledgments" to previously published work. (Informal communications may also be acknowledged in articles, but this is a different type of exchange and will be considered later.) Recognition is thus publicly awarded by individual scientists. It is usually necessary, even obligatory, for them to recognize previous work, for the validity of their own contributions depends logically on the earlier work. Recognition of this type, then, is concerned with the central values and norms of science: the only matters that need be considered are the value of the earlier work, its bearing on the problem under discussion, and its validity. The conformity of the original donor to the "peripheral" norms of science does not enter into consideration. For example, the original donor may frequently deviate from norms regarding acknowledgment of earlier work, or he may lack objectivity and refuse to consider contributions made by certain classes of persons, such as Jews or believers in indeterminism, but all this makes no difference. If his work is relevant to the problem under consideration and if it is valid, it must be recognized.

Over a period of time an individual may establish a reputation through his successive contributions. The word "reputation" usually means a good reputation, a reputation for contributing valuable and valid work, although presumably individuals may also develop reputations for contributing trivial or invalid work. Those with good reputations are more likely to have their contributions read and acknowledged. In most fields, because more articles are published than any one person can read, articles by eminent men are more likely to be read. For example, a mathematician said:

> If an obscure person has done something and then publishes it in an obscure journal, fewer people will look at it than if an outstanding person had done the same thing, because when people pick up a journal they scan the titles and look at the names of the authors . . . the work of certain authors will always be looked at. . . . It is also true that well-established people will have their work accepted quite readily, without too much refereeing.

Especially when work is published in abstract form or when full details of experimental procedures or mathematical proofs cannot be given, scientists often estimate the probable validity of the work on the basis of its author's reputation or, in some fields, on that of the organization in which the author works. Hence, those who have done good work in the past are more likely to be given credit for doing good work in the future, even if they are anticipated by others, whereas those who are little known are likely to remain so.

One theoretical physicist was writing an advanced text. It was suggested to him that the author of a text in a modern field must assign credit and that this might be a problem:

> An annoying problem. I decided originally to write the text in order to master the field, really learn about it. In doing this reading I have noticed that the current credit assignment in the field is unfair. Well-known physicists are likely to get credit for things even when others deserve more of it. One well-known theoretical physicist never gives credits to earlier work;

he justifies this by saying that it's very hard and takes a lot of time. Another theoretical physicist maintains that one should be very careful about this. I'm not excited by reading history and am partial to the first point of view; however, if I were to adopt it I would generate animosities. . . . Big steps forward in science are really built up out of a large number of small contributions. I know this is true with respect to quantum mechanics and relativity. Many things preceded the work of Einstein and Schrödinger; the preliminary work was already done and someone else would have done it if they hadn't.

A younger theoretical physicist expected this to happen to him. He and a colleague, also a young man, had written a paper applying a new theory to solid-state physics, a paper of which he felt quite proud. A similar paper was written about the same time by a Professor Z, a man with a wider reputation: "Z's paper and the one by X and myself are similar, but his is more likely to be recognized. Because his name is more well known, people are more likely to read his paper." He felt this was a typical experience.

Other scientists, to whom these examples were cited, denied the possibility that recognition would be unfairly apportioned in their own specialties.[42] A typical response was given by a mathematician:

In mathematics it is all very clear-cut. We judge work by the quality of the paper. A man can become well known overnight—that is, within a year of publication. . . . It is possible that if someone publishes topological results in a completely different aspect of mathematics it might take longer for his work to be recognized.

To some extent scientific specialties may vary in regard to the fairness with which work is recognized. However, the perceptions by scientists of the informal organization of science are often distorted, and informants equally well placed to observe behavior often make inconsistent reports. In the present instance, a belief that the system of recognition in one's field is unfair is clearly incompatible with a motivation to do good work in order to be recognized, so that distorted perceptions may lead to conformity. This is especially likely because recognition, as many other forms of informal prestige, is extremely difficult to evaluate. Sociologists, experts in the evaluation of prestige, sometimes find ways of doing so precisely; in science, for example, one could count references to a man's papers by others in their publications. But scientists themselves, unlike actors, seldom possess a clipping collection. Their own judgments of the recognition of their work are based on relatively few cues. (Among the elite in science, of course, many more cues are given.) A theoretical physicist felt that his best work (part of his doctoral dissertation) had been properly recognized, but when asked how he knew this, he said: "I don't know how much it deserves; the paper was published and duly considered by those who discuss these matters."[43] A mathematical statistician, a leader in his small subspecialty, was more hesitant to state an opinion:

> *What about the recognition due you? Is that fairly apportioned?* I think so, but it is hard to answer that question. I have very little idea of my own reputation, and I suppose that is true of most people.

Imitation is the sincerest form of flattery, and it is therefore a clearer ground for determining one's reputation and recognition. One example of this deserves to be quoted in detail to give the reader an idea of the process. The informant was an experimental physicist working on solids:

> I mapped out an experiment and arranged for a graduate student to construct the equipment and look for the resonance. Others had thought about such an experiment but expected the signal to be weak and were therefore discouraged. But we thought we had a marginal chance of finding it, in view of recent electronic developments. We built the equipment, checked it out—and found that it didn't have the required sensitivity. We were discouraged and almost ready to give it up. Then I got a flash, an inspiration. I saw that we had underestimated the strength of the signal by a factor of about 10,000. We just put our sample in and it worked. Other people were discouraged from looking for the effect because they had underestimated the strength of the signal. There is an anticlimax to all this: our general argument that the signal was enhanced was all right but not in its specifics. But we published a letter on it quickly. . . . Anyway, in the last year, since that came out, maybe a half-dozen laboratories here and abroad, including Japan especially, have begun to use the technique with other metals and alloys. It is now an expanding field. . . .

Even being emulated by others, however, is not always clear, nor is it always a good clue to one's reputation. For example, a molecular biologist said that his most important work was concerned with the metabolism of a certain microorganism. The work was apparently emulated and followed up by others, but it was not easy for him to be sure that it was *his* work which was being emulated: "Great work was being done at the same time with other microorganisms of a similar sort."

Errors may detract from a man's reputation. Sometimes this is a serious matter, although frequently errors are caught by referees and editors. A theoretical physicist was thankful for this check:

> A couple of times reviewers for the journals challenged my papers. In one case the reviewer was completely wrong, in the other case I was completely wrong. In the former case the criticisms only applied to part of the paper, but the reviewer was wrong, and I published. In the latter case I was completely wrong and thankful to the reviewer.

When this check is not present, as in unrefereed journals and papers delivered at some society meetings, serious consequences may occur. A physical chemist told me of one example:

> At society meetings almost any paper can be presented. The American Physical Society requires a two hundred-word abstract, The American Chemical Society 1000 words, but papers are evidently never turned down since so much poppycock is presented. Some damn good people are sorry things are not reviewed prior to presentation in society meetings. I know one case in which it cost a guy a position. He was a good man but had one of his poorer days, and the fact that his contribution was not reviewed meant that he presented something quite wrong.[44]

It is also true that referees may fail to catch an error, which may then be published. A mathematician was asked if this might be embarrassing to the author:

> Not only embarrassing, but it could be a serious blow to his reputation, particularly after he's pretty well established, to come out with a result for which he must apologize afterwards. After all, mathematics is supposed to be rigorous. . . . It is conceivable that the first mathematician may present his argument in such a plausible way that others follow him into the same trap and incorrect mathematics may be in the literature for a long time. There have been important cases of this, where untrue theorems not only get into the literature but into texts; this is serious.

It should not be concluded that scientists have a compulsive fear of error which leads them to emphasize accuracy at the cost of other scientific values. Different types of errors have different costs. Significant work will be recognized even in the presence of error or inelegance. The case of the solid-state physicist already quoted is one example of this. Other examples are found in mathematical fields. An eminent topologist said that he had published mistakes and that this was common. "Someone once observed that more reputations have been based on incorrect proofs than on correct ones."[45] Similarly, a brilliant mathematical statistician said:

> I have never had a paper that wasn't challenged. I've never published a paper without a serious mistake and I never expect to do so. . . . Usually this is handled through correspondence. Someone will write and say "Your proof of lemma so-and-so doesn't hold," or something like that. A famous editor of a mathematical journal said the same thing as I have about errors. Papers without serious mistakes are probably trivial—the work is too easy.

This is an extreme position. Scientists differ in their concern over technical errors, even the same kinds of error. Disciplines differ in the kinds of error likely to occur and in the degree to which errors harm an individual's reputation.

A final form of institutionalized recognition is the collective honor. Scientific societies and research institutions often honor the work of leading scientists by awarding them medals and prizes.[46] These institutions, as well as groups that publish scientific works, also frequently give leading scientists the opportunity to make contributions in ceremonial settings. For example, universities often have honorific lectures; scientific societies sponsor symposia at which eminent men are invited to read papers; and journals publish review articles by leading scientists. (The award of a prize or medal is often accompanied by a scientific lecture given in a highly ceremonial setting. This is true, for example, of the Nobel Prize. The combination of the award of honors and the donation of a ceremonial contribution of information is also exhibited in the practice of publishing volumes of papers in honor of distinguished scientists—*Festschriften* and commemorative volumes. In such ceremonies as the Nobel Prize award, the scientist receiving the award contributes the ceremonial information; in *Festschriften* the scientist is honored by contributions given by others in his honor.) These rewards are frequently made for specific contributions. The Nobel Prize in physics, for example, is given for "the most important discovery in the realm of physics" in the preceding year, the Prize in chemistry for "the most important chemical invention or improvement," and so forth.[47]

Other awards are given for less specific accomplishments. Authors to review articles, for example, are selected for their general leadership in the field, and such awards as the University of California Faculty Research Lecture are given to a man "who has distinguished himself by scholarly research in his chosen field of study." The recipients of such honors are usually selected by a small group of persons who act in the name of a wider scientific community. It is reasonable to suppose that these forms of recognition, unlike the citations of published work that appear in research publications, take into account not only the recipient's conformity to the central norms and values of science but his conformity to peripheral norms as well. That is, perhaps the scientist who constantly neglects to cite the work of his predecessors, who lacks objectivity, and who vilifies those he disagrees with is less likely to receive such awards, even if his results have been of central importance and valid. Those who make such awards are not obligated to neglect such peripheral considerations.[48]

Thus, there are various levels of formal recognition in science. An individual establishes his status as a scientist by having his research contributions accepted by a reputable journal; he achieves prestige as a scientist by having his work cited and emulated by others; and he achieves elite status by receiving collective honors. Collective honors have other important functions besides that of allocating elite status. One of these may be termed the "celebratory function." In conferring the honor, the scientific community celebrates what are felt to be its collective achievements. Such celebration tends to affirm the collective goals and the solidarity of the community.

ELEMENTARY RECOGNITION

Interpersonal approval and esteem take a wide variety of forms in science, greater than institutionalized recognition. The kinds of information exchanged and the general social functions also vary. However, a simple classification can be provided to give an idea of the range of behavior involved. I shall first describe the typical situations in which approval and information are exchanged and then discuss the general social functions.

The meetings of scientific societies serve as forums in which both institutionalized and elementary forms of recognition are awarded. The former has already been described; it was noted that reading papers at society meetings is a technically unsatisfactory way of transmitting information and, because of the small audience, an unsatisfactory way of achieving recognition. Scientists frequently assert that the major function of such meetings is not the reading and reception of papers but the chance to meet one's colleagues face to face. A typical statement was made by a solid-state physicist:

> Meetings are important. You don't go so much to hear papers as to talk to people. I have read papers at meetings, but the informal aspects are more important. Several meetings are focused on the informal aspects. . . .

Such smaller meetings were always evaluated more highly. They include the Gordon conferences[49] and meetings in special subfields organized by larger scientific societies. Eminent men may themselves organize informal meetings for longer periods of time during the summer months. Two of my informants had done so. One, a molecular biologist, has about thirty men in his specialty gather at his institution every summer. Another, a mathematician, had organized a special conference in his field for the summer; many of the participants were supported by the National Science Foundation, but others came at their own expense. Because these smaller and more informal conferences are usually open solely by invitation, only the more eminent men attend. Some of the larger society meetings, on the other hand, are attended by less eminent men, men from low-status organizations, and especially persons from industry. Since this is so, eminent men often disparage them; for example, the molecular biologist who organized his own informal conferences said, about the larger society meetings: "I think it is distracting and takes too much time. I really think that this society business is of very little use to science. . . ."

In addition to meeting informally in the context of more formal gatherings, scientific colleagues frequently visit one another at their home organizations. Such visits may be formally arranged as well, for example, the state visits of eminent scientists and the visits of hired consultants. When visits are not possible, scientists in a particular field often maintain an extensive research correspondence.

Finally, the most frequent form of informal contact is among colleagues in the same organizational unit, most importantly, the same university department. Here informal contact is often continuous, although, in many cases, the scientist will find no one else working in his own problem area among his departmental colleagues.

In such informal contacts, the information conveyed ranges from the very precise and specific to the most questionable and diffuse.[50] Informal contacts often serve as supplements to formal channels of communication or for trips to the library. For example, the exchange of "preprints" (duplicated copies of papers distributed before publication) supplements publication in journals. The scientist can disseminate news of his discoveries to his colleagues far more rapidly through the use of them, and this is often important in developing fields.[51] A mathematician in a leading department said, "If something important happens, the world of mathematics knows of it quickly through letters." Preprints are usually sent out to a list of those the author feels would be interested. It is, therefore, often desirable for the scientist to cultivate informal relations with others in order to receive their research results before publication. While many scientists do not distribute preprints, others have a lengthy mailing list of those to whom they send results. One mathematician sent his preprints to 105 colleagues around the world; he asserted that he had personally met them all, except for a few Japanese. The distribution of preprints (and reprints) is marginal between formal and informal channels of communication. Preprints may be a less "responsible" mode of communication. The author may ask his colleagues for suggestions and corrections before publishing the work.

At other times, the person who desires specific items of information seeks out the one he thinks can supply it. Sometimes a solution can be given quickly and immediately, especially in the mathematical sciences, or the informant will refer to books or articles in which the information can be found. Some individuals are better sources than others, but usually men in the same department are asked first. Information of this type is usually exchanged for approval and esteem. It may also be exchanged for rights to reciprocal assistance of the same type.[52]

Especially in experimental sciences, the information conveyed may be far more technical and require greater effort on the part of the donor. It may involve either instruction in the use of complex equipment or the actual use of it. When the services requested require relatively little time and effort, they may be granted and the donor may receive, in return, deference and rights to reciprocal services. When considerable time and effort are required, these may be insufficient rewards and the donor may claim a share in the recognition given the finally published work, viz., he may claim the right to be listed as a co-author on publications emerging from the work. Scientists resist being downgraded to technicians. If a person to whom they contribute services cannot himself supply them with the recognition the services call for, they will expect a share in the recognition the larger scientific community gives for the finished product. A theoretical physicist mentioned such expectations in saying: "I don't work effectively without direct and continual exchange of ideas. And I can't expect to waste someone's time talking unless he becomes professionally involved in the same problem."

Technical information is contributed more easily and therefore more readily than are technical services. Scientists expect to contribute information to those who seek it from them.[53] In many, perhaps in most, instances, the problem of being downgraded to a technician does not arise. Axel has argued:

> The most important function of exchange visits is the exchange of ideas and not of technical details. These ideas are often subtle and are not fully clarified until after an extended question and answer period. When a noted scientist visits, the entire "home team" can co-operate in learning what he knows; each local specialist can learn about his specialty. There is then a great advantage to an institution if it can arrange for experts to visit; both its mature scientists and its students have a valuable opportunity to learn.[54]

Furthermore, speculative material and certain kinds of polemical material are not permitted in many scientific journals. Therefore, such information can only be transmitted by way of correspondence or face-to-face meetings.

Formal channels of communication demand responsibility: the scientific article is expected to be a finished and polished piece of work. Informal channels of communication, on the other hand, often involve a great deal of permissiveness. People can make suggestions without committing themselves, and others can criticize the work without having to make any final decision with regard to its validity. A young theoretical physicist seemed especially dependent on such contacts:

> You get a crazy idea and go chasing to one of your friends and say, "What about this?", and he'll say "Crazy," and you'll bow your head and go home and think of another idea. And with the next one your friend may be interested. . . . In theoretical physics it is very hard to be right the first time by yourself. . . . You're never quite sure whether an idea is correct or whether it's really worth-while. There is always a temptation to think that all ideas you invent by yourself are good by definition. There has got to be a curb on this, I think. Intercourse at this level is extremely important. . . .
>
> We're always argumentative about everything. . . . The game is, you try to smash everybody else's theory. Somebody comes up with a theory,

> and you try to prove it's wrong, and if you can't prove it's wrong, then you start working on it. . . .

Similarly, a leader in the field of theoretical physics cautioned me to be skeptical about his comments on the social organization of science: "You should argue with me. I'm used to talking with other physicists, where, if I present wild ideas, they will be refuted by others." In such instances, the donor of the information is indebted to his critics, although, as indicated, the recipients of the information may themselves find it useful. The donor, however, shows his respect for his recipients by asking them to evaluate and criticize his own ideas.

Informal channels of communication also play a celebratory function. Acquaintances can congratulate one another directly on their successes; doing so defines the results as successes. This often induces conformity with the goals of the field; a mathematical statistician said the following about this:

> My contacts with others are more social than intellectual. . . . We talk to one another and criticize one another's papers. . . . The contacts are important to my work, very important. They keep you aware that what you're doing fits into the scheme of things. They influence the direction in which you're going—you don't want to do exactly what others are doing, but you also don't want to get too far from others. . . . You get recognition from others.

The contacts reinforce the motivation of participants. A mathematician said:

> If I have any regret in my voice in talking about how few I am in contact with, it is certainly this, that contacts of this sort are extremely valuable; in exchanging ideas you get new ideas yourself. It helps tremendously. I know I was all fired up because of this last conference. It's a little more difficult to have this enthusiasm if you're working by yourself. These conferences certainly tend to generate it.

Other scientists avoid meetings and conferences for this very reason. The inability to move toward the goals one assumes may be frustrating. For example, a relatively inactive experimental physicist seldom went to meetings:

> My experience with meetings is this: I come back with enthusiasm and ideas, but when I get here I am baffled by lack of funds and help, with no chance to do anything. Thus meetings are frustrating. I get the information through journals.

The social behavior of other scientists is often discussed in these informal meetings, and this plays a related function. Gossip is an effective form of social control, especially with regard to peripheral norms regarding good behavior. Since it involves criticism of the behavior of a person not present, gossip makes it possible to state a norm without exhorting persons present or criticizing them. The following kind of evaluation is commonly made in informal conversations among scientists. An informant responded to a question about an eminent mathematician's views on competition by saying:

> X is obsessed on this point; he has written a novel about it, I guess. He is really quite immature. He feels he hasn't received the credit due him for his work. I've met him, and he really fits the extreme things people say about him.

Finally, informal contacts may induce solidarity in organizations conducting research. Scientists in research laboratories and university departments must often make collective decisions in competitive situations, where rivalry and antagonisms are likely to arise. That members are interdependent in their research activities, or at any rate feel they owe deference to those who have assisted them at one time or another, helps maintain the solidarity of such units.

Some forms of both institutionalized and elementary channels of communication have now been described and analyzed. The discussion has dealt only with communication among colleagues. This stress is justified, for scientists themselves place most importance on communication with their colleagues, and the social control exercised through such communication is the most important social influence on the activities of scientists. For this to be so, however, the scientific community must have some influence over the communication between its members and nonscientists: extracollegial forms of recognition can threaten the integrity and autonomy of the scientific community.

EXTRACOLLEGIAL RECOGNITION

Most pure scientists play roles in which they are expected to inform nonprofessionals. These are typically students, but they may also be technicians or members of the general public.

University professors interact with graduate students far more than with undergraduates, and only the former will be considered at any length in this volume. A professor is responsible for conveying general information to his students in the form of lectures, and he also serves as research director to a small number of them. This relationship is important, and many professors discuss their research more often with graduate students than with anyone else. In addition to teaching the student, the professor is responsible for evaluating him and, to some extent, for getting him a job when he completes his work. These activities are not typically gifts. The professor is paid for them and may be formally obligated to accept students. Coercion of this type is not usually necessary, however, for the professor gains deference from students, and they often assist him in his research. The professor's conformity to higher scientific goals and values may be threatened if he becomes unduly dependent on the deference and esteem of his students. Students are, by definition and usually in fact, incapable of carefully evaluating his work. Furthermore, their deference is more or less obligatory, for the professor holds their fate in his hands. Because of this, the esteem of students is generally given far lower value than recognition by colleagues.

Scientists in universities, industry, and government are also expected to respond to requests for information from nonspecialists. These may be scientists in other disciplines, technologists, or members of the general public. The important characteristic of this, from the present perspective, is that donor and recipient do not share goals and values. Scientific information is utilized by the recipient for nonscientific purposes or, in interdisciplinary exchanges, for the goals of a different branch of science. Because the scientific donor does not share the goals of the recipient, approval and esteem by the recipient is not as valuable. Of course, it is often prized. For example, a statistician often gave information and advice to medical scientists without fee; when asked why, he answered:

> I get a lot of factual information. . . . And it does suggest problems for my own research. . . . And, again, in the consulting situation the person who comes to me suffers some ego loss, and this accrues to the consultant. These ego gratifications are a form of reward.

Despite these rewards, this informant felt that he ought to be paid for his consulting activities; he was chagrined because he was not paid by the nonuniversity agency for which he was a consultant, whereas physicians who acted as consultants for the same agency were paid $50 or more daily. Scientists generally expect pay for services to those outside their own disciplines. The information is not a gift, since the acceptance of it contributes neither to the donor's status as a scientist nor to his prestige within his discipline. Of course, disciplines differ in this regard. Statisticians often expect to work with nonstatisticians, and many of them regard the opportunity to do so as one of the nice things about the field. It is also probably true that a statistician's contributions to nonstatistical studies enhance his reputation as a statistician. (Most of the leading statisticians of this century have made significant contributions to other fields, from Karl Pearson's contributions to biometry and Sir Ronald Fisher's work on the genetics of evolution, to Jerzy Neyman's work on the distribution of galaxies.)

Although contributions to other disciplines, engineering or medicine, may enhance a scientist's reputation in his own discipline, the same is far less likely to happen as a result of contributions to the lay public. The esteem of the lay public, or the "great unwashed," as some experimental physicists have characterized them, can only weaken the scientific commitment of the man who becomes dependent on it. Consequently, popularizations sometimes appear to reduce a man's prestige within science.

The late Richard B. Goldschmidt, an eminent biologist who wrote many popular works, perceived this as a "strange fact":

> It is a strange fact that scholars frequently are strongly prejudiced against popular writing, thinking that it lowers the standards of a research scholar and makes him a superficial writer. The great examples of Huxley, Helmholtz, Haeckel, and Fabre are usually forgotten. Actually, in my opinion, it is the duty of the research man to write popular science. . . .[55]

Most popularizations, however, are written for money and not out of a sense of duty. Popular recognition is not enough. For example, an experimental physicist who had written many popular articles was asked if his colleagues appreciated it:

> I wouldn't want to overdo it, but in some areas we almost have a responsibility to do it. There has been a change in attitude over the past few years. I think this is largely due to the *Scientific American*. Besides, you get paid well for popular articles now. . . . The administration here likes a reasonable amount of popular work; it's good publicity for the institution.

Norbert Wiener wrote of his *Cybernetics* that it "represented the beginning of whatever [financial] security I enjoy at present."[56] As the physicist quoted indicated, popularizations may also contribute to the security of the institution in which the scientist is employed. A molecular biologist said that his department recruited students as a result of a TV series it produced. The Director of the Oak Ridge National Laboratory has suggested that the financial needs of "Big Science," represented in the large research institutions, may lead them to popularize their discoveries, to obtain political support for their budgets, and he suggests that this may have a corrupting effect on science.[57] Compare this statement by Samuel A. Goudsmit in *Physical Review Letters:*

> As a matter of courtesy to fellow physicists, it is customary for authors to see to it that releases to the public do not occur before the article appears in the scientific journal. Scientific discoveries are not the proper subject for newspaper scoops, and all media of mass communication should have equal opportunity for simultaneous access to the information. In the future, we may reject papers whose main contents have been published previously in the daily press. . . .[58]

The scientific community is jealous of its monopoly as the source of scientific recognition, for the source of recognition is in many respects the source of control over the direction of research.

THE ARTICULATION OF INSTITUTIONALIZED AND ELEMENTARY RECOGNITION

While institutionalized forms of recognition are most important in maintaining conformity to higher scientific norms, the elementary forms mediate between the larger scientific community and the individual scientist. This is often an important function, for institutionalized forms of recognition are in many ways unsatisfactory forms of gratification.[59] First, institutionalized recognition is accorded in the future, often long after the research is concluded. Its gratifications must be deferred. Second, it is uncertain, like the success of works of art or Broadway plays; it may even be posthumous. Third, such recognition is impersonal. It is awarded, as we have seen, for specific accomplishments, and it does not go much beyond them. Such recognition is not concerned with the individual's wider merit as a man. Fourth, and finally, institutionalized forms of recognition do not establish interpersonal rights and obligations.

In each of these respects interpersonal approval and esteem complement institutionalized recognition. Interpersonal esteem is often a direct source of gratification; the individual scientist and his close colleagues may almost immediately celebrate his successes, however uncertain they may be. Colleagues, in personal contacts, may help maintain each other's confidence in their research. This esteem and approval, being interpersonal, tends to be diffuse, pertaining not only to specific scientific accomplishments but to the larger personality of the discoverer. Finally, as we have seen, the elementary forms of recognition often establish bonds of deference and obligation between individuals; in such intercourse, the individual may establish rights for future services and gratifications.

Thus, the forms of recognition awarded in primary groups of scientists tend to make the institutional incentives meaningful for day-to-day work. They typically reinforce the effects of institutional incentives, and without them scientists might conform less to the norms and values of science: they might be less disposed to work, to publish, or to select problems and techniques within the scope of their disciplines.

However, as in any community, small groups may work against the norms of the larger community instead of reinforcing them. That is, recognition and status in the primary group or the small organization may become more important to the individual than recognition and status in the larger community. If the norms and values in such smaller groups differ from those of the larger community, commitment to the latter will be weakened. This may be the usual experience of scientists in industrial research. For example, a mathematician told why he left a position in industrial research:

> The people with real prestige were the administrators. There was more immediate gain for completion of short-term projects. . . . When I looked ahead and thought of what I wanted to do, I wanted to continue doing research in mathematics. I could see very few senior people in industry who had done that. . . . They became administrators, they became engineers, or they left.

Among scientists who remain in industry there is a strong tendency for the incentives offered by the employing organizations—interpersonal approval as well as formal status and salary—to become more important than recognition by the larger scientific community.[60] This might also occur in universities or basic research organizations.[61] Such organizations could develop norms and values that deviated from those of the larger scientific community; if the esteem of immediate colleagues were sufficiently important, scientists working in such organizations would tend to deviate from the goals and values of the larger community. However, this does not usually occur. Two reasons seem especially important. (1) Goals are seldom shared perfectly by the members of a university department. There is a tendency for them to work on somewhat different problems, in order to avoid intraunit competition. This diversity of goals is made more likely by the disposition of university departments to avoid inbreeding by making new appointments from among the graduates of other institutions.[62] (2) As is well known, organizational incentives in universities and other agencies conducting basic research are linked with the institutionalized recognition of the scientific community.[63] Salaries and promotions in such organizations tend to be based on the number of papers a man publishes and the prestige he has among his scientific colleagues. Meltzer, in his analysis of the results of a mail questionnaire returned by 75 per cent of all United States physiologists, found that those working in universities and government agencies were more likely to say that papers counted in promotion and more likely to publish than those employed in industrial organizations. Salary was correlated with number of publications in all three types of institutions. Caplow and McGee found

TABLE 2

THE IMPORTANCE OF PUBLICATIONS IN THREE TYPES OF RESEARCH ESTABLISHMENT

	TYPE OF INSTITUTION		
	Academic	Governmental	Industrial
Percentage reporting papers count to at least some extent in promotion	86	83	39
Percentage reporting five or more papers in a three-year period	54	59	31

Source: Leo Meltzer, "Scientific Productivity in Organizational Settings," *Journal of Social Issues*, 12 (1956), 37–38, tables 4 & 5.

that "disciplinary prestige is the most important measure of evaluation when the roster of candidates is being assembled" for any academic vacancy, although additional criteria may be used in deciding among individuals of sufficiently high prestige.[64] They also found: "The higher the prestige of a department, the greater will be the tendency for its members to be oriented to the discipline rather than to the university."[65]

Superior academic institutions place most emphasis on professional reputations and publications, and their scientists are responsible for most of the basic research published. This is indicated by a study of the authors of papers in the *Journal of the American Chemical Society*, in which roughly two-thirds of the fundamental research in chemistry by American chemists is published. Papers in the 1949 volume were classified, with the results shown in Table 3.[66] More industrial research is done in chemistry

TABLE 3

PLACE OF EMPLOYMENT AND PUBLICATIONS OF CHEMISTS

SECTOR	PERCENTAGE OF TOTAL PAPERS, J.A.C.S., 1949*	PERCENTAGE OF CHEMISTS WITH PH.D., 1948**
Academic laboratories	64	33
Industrial laboratories	19	54
Government laboratories	9	10
All others†	9	4
TOTAL	101	101
(Number of papers or chemists)	(1,086)	(574)

* Joseph A. Kraus, "The Present State of Academic Research," *Chemical and Engineering News*, 28 (September 18, 1950), 3203–3204.

** U. S. Bureau of Labor Statistics, *Occupational Mobility of Scientists* (Bulletin No. 1121, Washington: Government Printing Office, 1953), p. 8. See U.S. Bureau of Labor Statistics, *Education, Employment, and Earnings of American Men of Science* (Bulletin No. 1027, Washington: Government Printing Office, 1951), pp. 3–5, on sampling procedures. Nondoctorates are excluded, although they might contribute papers; including them would increase the proportion of chemists in industry and decrease the proportion in academic employment. Women, who comprised 5 per cent of U. S. chemists in 1948 (*ibid.*, p. 7), were also excluded from the table.

† This includes, for papers, foreign sources as well as other U. S. sources. For place of employment it includes self-employment, employment as independent consultants, or employment by nonprofit foundations.

than in any other scientific discipline. Academic laboratories produced a disproportionate number of all papers, and contributions from academic laboratories were concentrated in a few institutions. Contributions were received from chemists at ninety-eight different universities, yet four of them supplied 25 per cent of the total, twelve supplied 50 per cent, and twenty-eight supplied 75 per cent.[67]

The available evidence indicates that the universities, especially the better ones, not only do more basic research but do it better. In a *Fortune* survey of eighty-seven "outstanding young nonindustrial scientists," fewer than half thought that industrial laboratories contained *any* scientists equal in ability to the best scientists in universities.[68]

Thus, in the organizations in which pure research is done, the organizational incentives of salary and position are linked with the institutionalized recognition of the larger scientific community. In this way, the scientific community controls decisions of the organizations. Why should organizations act this way? To some extent, they may do so because they are directed by eminent scientists who are devoted to the values and interests of the scientific community. In addition, however, competitive exigencies compel organizations to link the incentives at their disposal with the system of institutionalized recognition in science. University departments compete for research grants, students, and staff. Research grants naturally tend to be allocated to the institutions in which the best research is done, and "best" here tends to mean "best in the eyes of the scientific community."[69] Students prefer institutions of high quality. They do so for good reasons, since students from such institutions usually get the best jobs.[70] Universities in which much research is done are able to choose better students, especially better graduate students.[71] As will be shown, faculties almost always want larger numbers of good graduate students, and universities actively compete for them.

The most important form of competition is for superior staff. Having a faculty with a distinguished research reputation makes it easier for a university to recruit students and to attract research funds. On the other hand, because scientists interested in basic research tend to prefer organizations in which good research is being done, they will often make financial sacrifices to be in such an environment.[72] Organizations that must keep research results secret have a more difficult time obtaining superior scientists. Even industrial research organizations, whose goals are only in the area of applied research, often appoint distinguished men to do basic research, give a small amount of time to other scientists for this purpose, and publicize their research results, solely to make themselves attractive to superior scientists. If they do not permit pure research, their applied research and development efforts suffer.[73]

Competitive exigencies therefore induce organizations to link their own incentives to those of the larger scientific community. It follows that, when such competition is absent, parochialism is more likely, and deviation from scientific goals and values will tend to result. Some evidence for this has been provided by Joseph Ben-David's comparison of the productivity of England, France, Germany, and the United States in the medical sciences from 1800 to 1925.[74] While England and France performed a large share of the research until about 1840, Germany contributed much in the last half of the nineteenth century, and the United States' contribution increased steadily from 1880. Ben-David first compares Germany with France. In many respects the countries were alike in their ways of favoring science, and in some respects France was ahead. In both countries the desirability of pure research was recognized; in both, standards of educational accomplishment were high; in both, educational systems encouraged research by advanced students. The significant difference lay in France's having a centralized system without competition between institutions, whereas Germany had a competitive system as a fortuitous result of political decentralization. In Germany:

> Improvements and innovations had to be made from time to time in order to attract famous men or keep them from leaving. In this way, laboratories and institutions were founded, assistantships provided, new disciplines recognized, and scientific jobs created. These innovations were repeated throughout the system because of pressure from scientists and students in general, irrespective of practical needs and of what a few scientific influentials thought.[75]

From the middle of the nineteenth century, British and American educators, scientists, and administrators displayed increasing interest in the organization of science in Germany. In both countries efforts were made to emulate German forms. This led, however, to more rapidly increasing productivity in the United States than in England. Again the explanation is to be found in differences in institutional competition in the two countries. British education, although not formally centralized, was dominated by Oxford and Cambridge. In U.S. education, there was competition as a result of political decentralization and the existence of both public and private institutions.

In the foregoing it has been shown that elementary forms of recognition mediate between individual scientists and the larger scientific community, that the goals and values of the latter tend to be supported, rather than subverted, by local groups and organizations, and that this is often a necessary response of the local groups and organizations. To those familiar with the U.S. academic scene, this may appear to have been belaboring the obvious. The emphasis on publication and professional reputation in American universities is commonly noted and often deplored.[76] It is suggested, with good reason, that teaching is neglected because of the emphasis placed on research. It is even possible that the administrators of universities and other institutions have assumed not only a responsibility for scientific research but authority over research workers as well. With a "publish or perish" philosophy, university administrators can compel faculty to publish. In many fields, those who control the allocation of research funds can control the selection of problems by scientists. Such pressures by administrators would have the effect of attenuating the control exercised by the larger scientific community through the reward of recognition. Where such administrative direction over research is operative, the system in which information is exchanged for recognition cannot be effective.

There are several reasons for believing that this has not occurred. It has already been noted that the allocation of research funds and, to some extent, the control of universities are in the hands of scientists themselves. University emphasis on publication does not seriously change this. The journals that accept scientific contributions are still organs of the larger scientific community. Furthermore, it appears that publication per se is insufficient; quality makes a difference.[77] One of the most important factors inhibiting the detailed control of administrators over research activities is the elaborate role-set of the scientist.[78] Most scientists are officially responsible for more than research, and formally, in fact, research may be a secondary responsibility, even if it is informally the most important. A survey of 42,000 "top scientists" (most of the sample had the Ph.D.) in the United States in 1948 found that only 23 per cent reported performing a single function alone. Usually research was formally combined with teaching, administration, or development.[79] The presence of the elaborate role-set apparently increases the importance of the colleague group. Other members of the role-set, such as administrators or those who utilize research in practical activities, are less involved. Scientific activities are carried on in contexts isolated from nonscientists. Although scientists can point to teaching activities as an important function, and thereby prevent the detailed direction of their activities by university administrators, students themselves have little control over the direction of research. (When students are not present, as in some foundation- and government-sponsored basic research institutes, it appears that administrators have relatively greater power because scientists cannot play off the claims of students against those of research supervisors.) Because most research is carried on in loosely organized contexts, information retains the nature of a gift from the scientist to the wider community, at least in a formal sense, and the most important source of control over the direction of research is the scientific community.

The elaborate role-set of the scientist may not only add to the power of the colleague community but also facilitate research more directly. The teaching role may produce better research scientists. The university professor must give lectures and, in so doing, must give his attention to the broader aspects of his field; this may help to keep his own research in the main stream of scientific thought. He will often be assisted by students. Unlike the assistants of the nonuniversity scientist, his assistants will have claims on him (to explain and justify his work to them), and they will be expected to make maximum contributions to his work. Veblen came to similar conclusions in 1918:

> Only in the most exceptional, not to say erratic, cases will good, consistent, sane and alert scientific work be carried forward through a course of years by any scientist without students, without loss or blunting of that intellectual initiative that makes the creative scientist. The work that can be done well in the absence of that stimulus and safeguarding that comes of the give and take between teacher and student is commonly such only as can without deterioration be reduced to a mechanically systematized task-work. . . .[80]

One of my informants, a theoretical physicist who had experience in both a university and an independent research institute, preferred the former. He liked close contact with students and experimental physicists.

> A good "experiment" for you is the theoretical physicists group at the [X Institute]. It is a failure, I think. First of all, there are no experimental physicists, and, second, there is no teaching. As a result, the theorists are faced with no small problems, problems suggested by experimenters or students, and they have no small daily successes. Hence their only goal is to solve some monster problem; since they can't always do this, they feel sort of guilty. Being in contact with experimenters and students, I don't have such problems. . . .

It is clear, of course, that some nonuniversity research institutes, and many scientists, have solved the problems mentioned by Veblen and my informant. But it also seems quite probable that teaching facilitates research.[81]

In the scientific community the institutionalized recognition of colleagues has precedence over the approval and esteem of those with whom the individual comes in contact, although such contacts are important for mediating between the individual and the larger scientific community. It does not follow that all scientists are predominantly oriented toward institutionalized recognition or that all scientists are partly dependent on the responses of immediate associates. All that has been demonstrated is the need for the predominance of institutionalized forms of recognition and the fact of this predominance in the larger community. Many scientists, even a majority, could place a higher value on the response of immediate colleagues without threatening the larger community, if the norms and values of the local group and the larger community reinforced one another. Similarly, large numbers of scientists may not require the gratifications of interpersonal response. Scientists differ in these respects because of the distinctive social positions they occupy and because of differences in personalities.

Individual Differences in Communication Practices

Productive scientists contribute information to their colleagues in widely varying ways. Such differences are made greater by the formal freedom most scientists possess; for example, they are not required to involve themselves in professional societies. Although each scientist is unique in the relative emphasis he gives to different forms of communication, a simple classification of scientists can be constructed on the basis of the preceding analysis of communication channels.

Consider the different degrees to which scientists may participate in various channels of communication, such as articles or informal contacts within university departments. The degree of participation can be crudely measured. One can, for example, count the pages a man contributes to learned journals, or the number of papers he produces, and one can also count the number of citations to his work in the same journals, the frequency with which his work is emulated, and the formal honors he receives. Within a department, one can observe how frequently a scientist sees his colleagues, the number of colleagues with whom he discusses research, and whether he is usually the one who initiates such contacts. The measurement of such behavior is difficult; details of the procedures used in this study are provided in the appendix to this chapter. For my purposes it will be sufficient to refer merely to "high" and "low" usage of communication channels.

The following channels of communication have already been discussed: (1) Published articles and books, and papers read at society meetings. This is the most important channel of communication from the standpoint of the larger community. Those who do not contribute at all through this channel cannot be considered scientists. (2) Contacts through meetings of societies. (3) Informal contacts with others in the same specialty at different institutions. These may occur through correspondence, visits, or in the course of scientific meetings. (4) Informal contacts with departmental colleagues. (5) Contacts with former and current graduate students. (6) Contacts with members of different disciplines or with nonscientists.

The set of persons with whom most communication is maintained will probably have the greatest influence on the scientist's own perspective.[82] Communication requires assuming the perspective of one's audience, and this eventually means making the perspective one's own. (Obviously this does not hold generally; a confidence man is not likely to assume the perspective of his mark. The norm of service in the professions and the scientist's typical desire for consensus make it more likely that performer and audience will share the same perspective.) Thus, the scientist who seldom communicates with his departmental colleagues but frequently has informal contact with those in his specialty at other institutions is especially likely to take the perspective of the specialist community, the scientist who disproportionately communicates with his students may take their perspective more often, and so forth.

With the above list of communication channels, a "profile" could be constructed for any scientist on the basis of his participation in them. If, for simplicity, involvement in each channel is dichotomized as "high" or "low," there will be sixty-four types (not all of which will be exhibited by actual scientists). Let us briefly consider some of the most characteristic and some of the most extreme.

(1) *Highly involved leaders.* These leaders participate a great deal in all the communication channels within the scientific discipline. They publish a great deal, receive formal recognition, participate in society activities, correspond, visit others and are visited by them, and spend much time in discussions with their departmental colleagues.

These activities require considerable time, and a leading mathematician said that as a consequence he did most of his research at home: "I don't do much research at the University but talk, teach, go to meetings, and so forth. . . . On the average, I spend four full days per week at the University, in classes, department meetings, colloquia, lectures, and so forth." This man is a leading expert in a rapidly expanding specialty shared by eight of his departmental colleagues. He has organized an informal summer conference for his specialty and exchanges many visits and much correspondence with other mathematicians. He produces many articles, but has only two graduate research students under his supervision (and considers that too many).

(2) *Informal leaders.* Some highly productive and respected scientists have many informal contacts but few formal ones. While they visit and correspond with others in their specialties and work in departments with others on similar problems, they avoid the formal activities of scientific societies. The distinguishing mark of these individuals is their tendency not to read the literature in their fields: informal contacts are used instead. Enrico Fermi, the Nobel Prize-winning physicist, was one such scientist:

> Already at that time (1928) Fermi made little use of books. . . . If he needed some complicated equation to be found in a book in the library, Fermi would often propose a wager, saying that he would derive the equation faster than we could find it in a book. Usually he won. . . . In this period [from 1945 to his death in 1954] we also note a change in Fermi's methods of keeping up with scientific developments. He read less and less and relied more and more on conversation and oral sources of information which were always plentiful. Many active physicists enjoyed and profited from discussing their problems with him and on his side he took notes of these conversations. . . . On the other hand, he barely glanced at the journals and completely stopped reading any physics books. He once said that Weyl's book on group theory and quantum mechanics was the last physics book he read.[83]

Two eminent physicists interviewed for this study possessed similar, though less extreme, habits. An eminent mathematical statistician summarized the attitude:

> I don't try to keep up with the literature. I think a man must decide whether to read or to write, and I choose to write. . . . I like to work with others. If I get an idea I try to interest others in the department in it; similarly, others will try to interest me in their ideas. More than half of my papers are jointly authored. . . . Maybe my collaborative work is why I don't have to search the literature. For example, the other day I wrote out a theorem I thought of on the blackboard of the coffee room. Now I'm waiting for someone to come along with a proof for it. . . . I find it more pleasant to work with others. . . .

This man, as the physicists mentioned, often visited others and received visits from others. He not only collaborated with members of his own department but with men in other departments of his university and at other universities.

Such communicative activity is only possible for eminent men in leading departments. They need not rely on formal communication channels because they associate informally with others who can supply the information they require. Others will seek them out and respond readily to their needs for information because they, in turn, can contribute to the needs of others. Scientists with similar orientations but without the eminence (and, possibly, the abilities) must rely on their departmental colleagues.

(3) *Scientific statesmen.* Men with established reputations in their own disciplines may devote much of their time to specialists in other fields and to nonscientists. Their prestige is secure enough for them to appear as experts to nonspecialists without jeopardizing their reputations among their colleagues. For example, a mathematical scientist had led something of a revolution in his field in the nineteen thirties. He was a leader in scientific societies, federal agency advisory committees, and scientific journals. Much of his recent work has been devoted to applying his specialty to other scientific fields. Another man, an experimental physicist with a respectable background, was also active with federal agencies, with scientific societies, and with a subject marginal to physics (astronomy). Scientists of this type have made contributions to their own field in the past and now contribute outside of it, while having relatively fewer informal contacts within it.

(4) *Student-oriented leaders.* Some eminent scientists noted for their formal contributions nevertheless spend a disproportionate amount of their time with their students. They may have little contact with their departmental colleagues. Although they may participate now and then in scientific societies, and while they may have some informal contacts with scientists working on similar problems in other universities, they relate to their respective disciplines primarily through their students. They often retain contact with former students. Sometimes one of them is felt to be the leader of a "school" consisting of his present and former students, and his eminence will stem partly from the success of his students in advancing his distinctive point of view.

(5) *Student-oriented scientists.* For a less eminent man, a group of present and former students may be nearly his only link with the scientific community. One informant said that some mathematicians are noted primarily not for their own work but for the work of their students, a description that seems to fit one mathematician interviewed for this study. He was relatively productive, yet most of his informal relationships were with students rather than departmental colleagues or colleagues in other institutions. He thought his situation was fairly common:

> Other members of the faculty in this department are working in this specialty but not in this subarea. Two men are working in a fairly close area, but we still have to study each other's papers closely to follow them. Most mathematicians have closer contact with their graduate students than with their professional colleagues. A mature graduate student can follow my work relatively easily. . . . I have contact with some older staff members and also with many former students of mine who are continuing to do research in this area. These contacts are extremely important to me.

This mathematician had written many papers with former students, although this is a relatively infrequent practice among mathematicians.

(6) *Intradepartmentally oriented scientists.* Some scientists, having strong needs for interpersonal approval and esteem but few students, and lacking the prestige necessary to approach specialists outside their own departments with confidence, must rely on their departmental colleagues. For example, one logician had written relatively little, and most of his papers were the result of collaboration with members of his own department. He had no students working on dissertations under his supervision. This man depended on others for assistance in publishing research. If, in the future, he proves unable to reciprocate the assistance others have profferred him, and if his professional reputation does not grow, it seems unlikely that his colleagues will continue to collaborate with him; his deference alone will not be that valuable. In that event his already strong orientation toward the instruction of undergraduates may become the dominant motive in his career.

(7) *Productive isolates.* Some scientists can continue to be highly productive while remaining relatively isolated from informal contact with their colleagues or with nonscientists, as were two of the mathematicians interviewed. That both were ethnically marginal does not seem crucial, since ethnically marginal scientists often have many informal contacts. Perhaps it is more significant that both were the only individuals in their departments working in their particular specialties. This situational aspect, combined with their personalities, accounted for their informal isolation. One of them said: "I could get along all right without such [informal] contacts. I am one who prefers reading to listening, writing to speaking."

(8) *Nonproductive isolates.* Naturally, some scientists may seldom communicate in any way with other scientists. When such isolation is sufficiently intense, the individual concerned is effectively retired from scientific life; he may turn his interests to the teaching of undergraduates.[84]

(9) *Marginal scientists.* The scientific statesman can devote much time to communicating with nonspecialists without jeopardizing a reputation he has already gained. The scientist who has made few formal communications to his own specialty, however, will tend to become a participant of the group he assists if he spends a disproportionate amount of time with them. For example, a statistician had been relatively unsuccessful in solving the statistical problems he selected, hence he had published little. (Lack of success may sometimes result from social isolation. This may have been the difficulty here.) His self-esteem apparently depended more on the recognition awarded him by some applied scientists he served as a consultant. Some time after he was interviewed, he left his university department to become an employee of the applied science agency where he had been a consultant.

In sum, then, the differences in the personalities of scientists and in the contexts in which they work permit wide variations in communication practices. To some extent, the variation can be viewed as a form of differentiation that helps meet organizational and institutional requirements. Science requires leaders who are willing to engage in organizational action even when it interferes with their own research; it is benefited by permitting creative geniuses to be influential without burdening them with organizational activities; it requires devoted teachers, and it needs eminent men to represent it in the wider society. A university department would probably lack solidarity if all its members were primarily oriented to the larger scientific community, and scientists whose social interests are centered within it can help compensate for this weakness. It is sometimes alleged that the procedures for selecting organizational leaders are haphazard and lead to precisely the reverse situation, viz., that "incompetent" scientists assume positions of organizational power.[85] In general, however, collegial bodies in science are organized "democratically," and men with distinguished reputations are given positions of power in them. This tendency for eminence to be accompanied by organizational leadership is but one example of the way in which various types of social relations are related in science.

INTERDEPENDENCE OF INVOLVEMENT IN DIFFERENT COMMUNICATION CHANNELS

The tendency for different forms of communicative practice to occur together can be seen most clearly in extreme cases. The men who lead scientific societies, edit scientific journals, and have close contacts with workers in a particular specialty throughout the world are likely to be those who have made many important contributions through published articles.[86]

At the other extreme, it is only the unusual scientist who can continue producing in social isolation. (The two productive isolates already referred to were more isolated from social contacts with other scientists than most of those interviewed, but at the same time they were in leading departments of mathematics and could not avoid some contacts.) By and large, isolation is associated with low productivity. Some of the reasons for this are shown clearly in autobiographical accounts. Leopold Infeld, for example, was prevented from participating socially in the activities of theoretical physicists largely because of anti-Semitic tendencies in Polish universities in the years following World War I. Of his position at that time he wrote:

> Scientific work needs encouragement, the presence of people with whom problems can be discussed exhaustively. With the exception of my short stay in Berlin, when I was too immature for research, I had never lived in a scientifically stimulating atmosphere. But I fully sensed its importance. I felt clearly that if one is not a genius the complete lack of scientific contacts must kill all desire to work. There was one danger more. In the isolation of Warsaw it was most difficult to work on vital problems on which groups of physicists worked with methods they created and developed. Like the Jewish Torah, which was taught from mouth to mouth for generations before being written down, ideas in physics are discussed, presented at meetings, tried out and known to the inner circle of physicists working in great centers long before they are published in papers and books. . . . I did not really believe that I should be able to continue for long any research in this atmosphere of isolation.[87]

Physicists interviewed for this study confirmed Infeld's view. One of them, trained as an undergraduate in a small European country, pointed out that scientists in such countries are often less productive because of their isolation. Norbert Wiener has given a similar explanation for the tendency of isolates to be less productive:

> I know of a number of cases where the relative paucity of scientific contacts, while not absolutely fatal, was still damaging and limiting. A scientist must know what is being done in order that the very individuality of his own work may come to full fruition. He must live in a world where science is a career, where he has companions with whom he can talk and in contact with whom he may bring out his own vein.[88]

Examination of the apparent exceptions to the rule that isolation is associated with low productivity tends to confirm it. There is, for example, the case of the brilliant Indian number theorist, S. Ramanujan. Solely with the aid of a rather elementary and obsolete text, he was able to achieve many of the results obtained by leading European mathematicians in the preceding century, and in many instances he went far beyond them. Until he corresponded with the Cambridge mathematician G. H. Hardy, he had no contacts, formal or informal, with practicing mathematicians.[89] Yet the only reason that Ramanujan's discoveries became contributions to mathematical knowledge was that he did come into informal social contact with Hardy and other Cambridge mathematicians. This informal mediation between the innovating mathematician and the larger mathematical community was a critical link in establishing his formal contributions and formal recognition.

Again, Albert Einstein tended to brag about his own isolation as a young man and suggested that isolation was often a good thing for scientists. He asserted, "Until I was thirty I never saw a real theoretical physicist."[90] He once seriously suggested that lighthouse keeping would be a good job for scientists who were refugees from Naziism.[91] But, as one physicist pointed out, Einstein attended many conferences; moreover, he was a student of H. Minkowski, whose geometry he later used in the general theory of relativity.[92]

Of course, it may be true that some isolation at some times is beneficial to the scientist. As one informant put it: "Some people argue that too much communication is bad, since it prevents one from developing his own ideas and results in his getting caught up in fads. Some isolation might be a good thing." But, in general, isolation means that the scientist will receive less recognition for his work and that he will lack information necessary for its completion. Isolation reduces the motivation to produce, especially to produce along the right lines, and it limits the capacity to produce by limiting the scientist's information. On the basis of the current data, it is not possible to decide whether the reduction of motivation or the limitation of information is more important. In general, both are important; the system of written contributions and formal recognition usually needs to be supplemented by informal contributions and interpersonal recognition.

The case for the association of various forms of communication can be based on more than the sketchy biographical data already cited here. The sample of scientists interviewed for the present study was small, seventy-nine, and the same questions were not asked in all cases; despite this, some suggestive results appear. The sample, or most of it, was classified as relatively high or low on six types of communication practices or related behavior.[93]

(1) *Participation in activities of scientific societies and similar groups.* Those classified "high" on participation had served, or were serving, as society officers, journal editors, or on advisory committees of grant-giving agencies. Others are grouped together, whether they usually attend society meetings or not.

(2) *Extradepartmental communication.* This was measured by the scientific correspondence and the amount of informal contact each scientist had with others in the same discipline who were in different departments.

(3) *Productivity.* This measure is based on the number of papers produced by the respondent in the preceding three years.

(4) *Special honors.* The scientists who had received special prizes, awards, or distinctive fellowships were contrasted with those who had not.

(5) *Intradepartmental communication.* This was measured by the amount of time a scientist spent in communication with departmental colleagues and the number of colleagues with whom he discussed research.

(6) *Number of students and postdoctoral fellows.*

Each of these measures was dichotomized. Table 4 shows the association (Yule's Q) between each pair of them.

TABLE 4

Correlations between Types of Communication Practice and Recognition (Yule's Q)*

	1.	2.	3.	4.	5.
1. Society participation	—				
2. Extradepartmental communication	1.00	—			
3. Productivity (articles)	.48	.85	—		
4. Honors	.60	.41	.62	—	
5. Intradepartmental communication	.55	.54	.42	.29	—
6. Number of students	.50	.58	.24	.12	−.01

* Yule's Q will be used throughout this work as a measure of correlation for double dichotomies. It ranges from −1, for perfect negative correlation, to 0 when the attributes are independent, to +1 for perfect positive correlation. When a Q is reported in the text, the number of cases, N, in the table on which it is based will accompany it. In this table, the number of cases on which the correlation is based range from 46 to 70; most correlations are based on more than 60 cases.

With one exception, all the correlations are positive, and most are fairly large. Productivity has a large correlation (0.85) with extradepartmental communication and a considerably smaller correlation with intradepartmental communication[94] and number of students. Since precise measurements could not be made (the same questions were not asked of each informant, and many answers were vague), some correlations may be spuriously high; e.g., men who had received high honors might have been judged to be more productive, regardless of the number of articles they had written. However, such errors seldom occurred, and in general the correlations adequately represent the communication practices of this sample of scientists.

It was impossible to separate the effects of different variables because of the small number of scientists involved. Donald C. Pelz made a study of a large government organization for research in the medical sciences in which the communication behavior of some three hundred scientists was analyzed in detail.[95] He was able to consider a number of variables simultaneously, and his results indicate some of the complexities in the relationships among different types of communication in science. Above all, however, it should be noted that he found isolates to have lower productivity on the average:

> Most scientists report average contacts weekly or more often with their organizational colleagues. Among the few isolates who report less than weekly contacts, performance tends to be low.[96]

Pelz was mainly concerned with the determinants of "performance." Judgments of each scientist's performance were obtained from his organizational peers and superiors. This measure is associated with the number and quality of the scientist's articles and books, since the organization studied is noted for its fundamental research and its goals and values are those of the larger scientific community.[97] It was found that the relation between a scientist's performance and his frequency of interaction with others depended on their similarity of values and background. Performance was highest when a scientist had daily contact with several colleagues who came from *different* types of institutions (e.g., one having worked previously in a university and the other in an industrial laboratory) and when they had *different* orientations to the institution in which they worked and to the scientific community. At the same time, frequent contact with at least one important colleague who had similar professional values was associated with high performance.

In the light of the description of scientific communication channels presented here, the most important of these findings is that high performance, or participation in formal communication channels, is *not* associated with high rates of participation in informal communications with other members of the same specialty when they are in the same organization. It is difficult to gauge the extent to which these findings could be replicated for university scientists.[98] For one thing, university scientists are almost completely autonomous. The role of the "chief" is absent. Furthermore, "similarity" of values and backgrounds will have a different meaning in the university context, if, indeed, it has any. On the one hand, disciplinary barriers are often greater in universities than in governmental research organizations: university scientists have less opportunity to meet with those in different fields. On the other hand, research in any particular university department tends to be highly variegated, with few individuals working on identical problems. In governmental research organizations, general goals will often be set by those higher up in the organization, and the possibility of coming into contact with those working on almost identical problems may be greater. Thus, university scientists will usually have greater opportunity to come into contact with others working on slightly different problems and less opportunity to come into contact with those working on greatly different problems. As a result, similarity of values and fields among colleagues in university departments has a different meaning than similarity among colleagues in governmental organizations.

The studies here summarized show that a scientist's uses of various channels of communication are not associated with one another in any simple way. It is clear that high productivity may be affected by informal contacts in different ways, depending on the context of the research and the individual scientist's status and personality.

Summary: Problems in Demonstrating the Consequences of Recognition

The thesis presented here is that social control in science is exercised in an exchange system, a system wherein gifts of information are exchanged for recognition from scientific colleagues.[99] Because scientists desire recognition, they conform to the goals and norms of the scientific community. Such control reinforces and complements the socialization process in science. It is partly dependent on the socialization of persons to become sensitive to the responses of their colleagues. By rewarding conformity, this exchange system reinforces commitment to the higher goals and norms of the scientific community, and it induces flexibility with regard to specific goals and norms. The very denial by scientists of the importance of recognition as an incentive can be seen to involve commitments to higher norms, including an orientation to a scientific community extending beyond any particular collection of contemporaries.

At the inception of this research, it was thought possible to test the exchange theory almost directly. As has been indicated, scientists can more or less order the problems open to them by their importance in relation to other outstanding problems, and they can similarly order techniques by the degree to which they approach ideals of precision, and so forth. This rank-ordering of problems and techniques will be more or less common to members of a discipline, and scientists oriented to recognition from their colleagues may tend to select important problems and valued techniques. As a student said:

> Mathematicians want their work appreciated by others, especially by other mathematicians. You don't generally do things which won't be appreciated. One rule for usefulness is to get results which are useful in terms of other mathematical work.

If one obtained a list of alternative problems in a discipline and their relative importance in the eyes of scientists in that discipline, one might be able to test such hypotheses as these: (1) scientists with the highest prestige tend to have solved problems in areas considered important by their colleagues; and (2) when scientists shift from one problem area to another within a single discipline, they tend to shift from areas of low prestige to areas of higher prestige.[100] If these two hypotheses were confirmed, the exchange theory would receive almost direct support.

Unfortunately, it proved impossible to obtain the data with which to test these hypotheses. In the first place, scientists were reluctant to discuss the importance attached by their colleagues to their problems. For example, a physical chemist was asked, "[Is your] research topic of central or of peripheral interest among physical chemists?" He replied, "I have never given it a thought." Another physical chemist replied to a subtler version of this question ("Do you think more physical chemists should be interested in your problem?") with, "I couldn't say." Such responses are to be expected if the prestige of his specialty is important to a scientist, for it would amount to asking him about his own prestige, and for the most part people resist answering questions of this kind. Of course, it was often possible to determine the relative perceived importance of different fields. For example, solid-state physicists generally agreed that solid-state physics is considered less important than nuclear physics, and mathematicians generally agreed that algebraic topology has high prestige among American mathematicians. But even here perceptions were often biased by the context in which the informant worked. For example, mathematicians working in departments where analysis is given much importance were more likely than others to mention functional analysis as a field of central interest to mathematicians generally.

In the same way, it is difficult to obtain good ratings of the prestige of individuals from a small sample of scientists in a few departments. However, if larger samples of respondents were presented with fixed lists of fields and individuals, one might obtain the data with which to test these two hypotheses. This would provide further confirmation of the information-recognition exchange theory of scientific organization.

At present, most of the evidence in support of the theory presented here is necessarily indirect. Such indirect support may consist of the rejection of competing theories or the demonstration of propositions more or less implied by the theory. The primary points of view—I shall call them "theories," although they have not been sufficiently elaborated to justify the term—that compete with the information-exchange theory stress either socialization or the importance of material rewards. The naïve individualism that denies the necessity for any process of social control has already been criticized. The other competing theory is more or less its opposite. Whereas naïve individualism holds that scientists are not motivated by a desire for extrinsic rewards of any kind, the other theory is that scientists are motivated by the same kinds of extrinsic rewards as "everybody else," namely, position and money. This would assert that the decisions of scientists are determined by the authorities who control these rewards—that scientists publish and work on certain topics and with certain techniques rather than others because only if they do so will they be rewarded by the higher authorities. This could be called a "contractual" theory of the organization of science.

Science is, in fact, an occupation and scientists do receive material rewards, and these are undoubtedly important. The problem for the information-recognition exchange theory is to determine if recognition and material rewards are actually consistent, and, if so, why. In this chapter, evidence has been provided in support of the view that the two types of rewards tend to be consistent and that material rewards tend to reinforce the operation of a system in which information is exchanged for recognition. The contractual theory contradicts this in two not entirely consistent ways. First, it asserts that material rewards are more important than recognition, that, for example, recognition is more likely to follow appointment to a distinguished faculty than to precede it. Second, it implies that this is the only way in which a system could be organized—although this implication is seldom made explicit. The theory appears to be mistaken in both respects. The most obvious evidence is the fact that many scientists who are under no coercive pressures to publish do so anyway. This is true, for example, of professors with tenure in large universities and of scientists in industrial research laboratories.[101] Furthermore, as has been noted, organizational pressures on scientists to publish are more likely to reinforce the power of the scientific community than to supplant it. There is little evidence that university officials have much influence over the choices of problems and methods of university scientists. The power of the colleague community is also reinforced by the elaborate role-set possessed by most scientists.

The contractual theory of scientific organization is an incomplete theory. It may account for publications, but it does not account for the selection of problems or methods. It might be argued that agencies which allocate research funds do this. However, evidence has been presented that scientists often control grant-giving agencies, so that the agencies tend to use the same criteria for evaluating scientists as do scientific colleagues when they award recognition. The contractual theory, unable to explain how "good" problems are selected and "bad" ones avoided, often leads to the conclusion that much published research is trivial and slipshod. This is, in fact, stated by some of its proponents, for example, Caplow and McGee:

> The multiplication of specious or trivial research has some tendency to contaminate the academic atmosphere and to bring knowledge itself into

disrepute. The empty rituals of research come to be practiced with particular zeal in unsuitable fields. . . .[102]

In England, the desire to expand the universities, and the consequent need to train less well prepared students with less well prepared teachers, has resulted in criticism of the system of the control of university scientists by the scientific community. Sir Eric Ashby has expressed some typical sentiments:

[The] young man who is inspired to devote his career to the real purpose of a university, which is teaching at the frontiers of knowledge, finds himself obliged to enter a different career: the rat-race to publish. And to publish what? It must be 'original': miniscule analyses of kitchen accounts in a medieval convent; the structure of beetles' wings—some beetle whose wings have not been studied before; the domestic life of an obscure Victorian poet; the respiration cycle of duckweeds. All, no doubt, interesting; all, in a way, at the frontiers of knowledge, even though it is crawling along the frontier with a hand-lens; all original, in the sense that no one has done them before; but all (with some few exceptions) so secondary to the prime purpose of a university.[103]

Such sentiments may be true. The problems of the quality of teaching will not be discussed here,[104] but another implication of the two preceding quotations should be emphasized, namely, that the contractual theory of the organization of science accounts not so much for its *organization* as for its *disorganization.*

Although this theory appears to be mistaken, it has not been shown that it is *necessarily* mistaken. Conceivably, the large government research organizations of the U.S. and U.S.S.R., or the large industrial laboratories here or in Western Europe, have developed or may develop effective science in which the autonomy of individual scientists is sharply restricted.[105] However, industrial and governmental basic research in the U.S., when it is effective, seems to be relatively unplanned and informally organized.[106] Such organizations often pride themselves on their "academic" spirit and rules. This may be necessary. The most important scientific innovations can seldom be predicted or obtained according to plan, and the acceptance or rejection of theories and methods must inevitably be left in the hands of the experts—not only individual experts but experts as a community of colleagues.

Additional indirect evidence for the information-recognition exchange theory will be provided in following chapters, beginning in Chapter II with the point that the scarcity of recognition leads inevitably to competition for it. If this were not so, the award of recognition would be ineffective as a means of social control. The consequences of competition will be discussed, along with some mechanisms for controlling undesirable consequences. The description of competition will be followed by analyses of scientific teamwork, the organization of subcommunities, and forms of disorganization. These analyses will illustrate the power of the information-recognition exchange theory of scientific organization.

APPENDIX. *Measures of Communication and Recognition*

This study has relied on each scientist's evaluation of his own communication practices and his productivity. Although this source of data is adequate for an exploratory study, it is subject to bias. Observation of overt behavior and an examination of published manuscripts would lead to more accurate measurements. The following variables are included in the tables presented in this chapter and in those to follow.

(1) *Participation in activities of scientific societies, journals, and on advisory committees of grant-giving agencies.* Because of the small number of cases, the sample was divided between those who were officials in such groups and those who were not. Since the information was not obtained for some scientists, and since most of the sample were not officials, correlations of this measure with others may be less reliable than other correlations.

(2) *Extradepartmental communication.* Scientists were asked how often they corresponded with colleagues on matters related to research and how often they discussed research with others in the same discipline who were in different departments. Their estimates, which were often vague, were classified as relatively high, medium, or low on this variable. In computing correlations, the medium and low groups were combined to divide the sample into two groups.

(3) *Productivity.* This measure is based on each respondent's count of his papers produced in the preceding three years. Those high on this variable have six papers or more, those with moderate productivity have from three to five papers, and those low on the variable have fewer than three papers. A scholarly monograph was counted as five papers, a text as one paper. Those who had received the doctorate very recently were not given productivity scores. Discipline is strongly confounded with this measure, as Table 5 shows. High rates of publication seem to characterize molecular biologists generally, not only those included in this sample; similar results were obtained by Berelson. For this reason, measures of association between productivity and other variables must be treated with caution. Another source of bias is the tendency for some scientists to include essentially the same work twice—once as an abstract and again as a completed paper. An attempt was made to omit abstracts, but some of them may have been included by respondents anyway.[107]

Some tables in the following chapters are based on a secondary analysis

TABLE 5

PRODUCTIVITY BY SCIENTIFIC FIELD

PRODUCTIVITY	FORMAL SCIENTISTS	PHYSICAL SCIENTISTS	MOLECULAR BIOLOGISTS
High	58%	54%	83%
Moderate	23	27	8
Low	19	19	8
TOTAL	100	100	99
(Number of cases)	(31)	(26)	(12)

of Berelson's data. His sample was large enough to make it possible to take scientific field into account in measuring productivity. Scientists in each general field were divided into quintiles in terms of the number of articles they had published in the preceding five years. The number of articles for each quintile is indicated in Table 6.

TABLE 6

PRODUCTIVITY BY SCIENTIFIC FIELD (BERELSON)

	NUMBER OF ARTICLES IN PRECEDING FIVE YEARS				
Field	*Fifth Quintile*	*Fourth Quintile*	*Third Quintile*	*Second Quintile*	*First Quintile*
Physical and mathematical sciences	15–51	9–14	6–8	4–5	0–3
Biological sciences	19–50	11–18	8–10	5–7	0–4
Socia sciences	11–99	7–10	5–6	3–4	0–2
Engineering	11–80	6–10	5	3–4	0–2
Humanities	8–50	6–7	4–5	2–3	0–1

Source: Secondary analysis of Berelson data.

(4) *Special honors.* Thirty of the seventy-nine scientists interviewed had been awarded some special honor by their colleagues. These included three Nobel-Prize-winners, seven who had received other prizes or honorary degrees, and seventeen who had received such distinctive fellowships as Guggenheim fellowships, membership in the Institute for Advanced Studies, and National Research Council fellowships. (Most postdoctoral fellowships were not counted.) In six cases it was not known if

such honors had been received; two of the six were classified among those receiving special honors on the basis of their known eminence.

(5) *Intradepartmental communication.* Scientists were asked how often and with how many departmental colleagues they discussed research. Again, because of the imprecision of the responses, they were more or less crudely classified as high, medium, or low on this variable. The medium and low groups were combined, to divide the sample into two groups.

(6) *Number of students and postdoctoral fellows.* Discipline is also confounded with this variable, as Table 7 shows. The distribution of

TABLE 7

NUMBER OF STUDENTS BY SCIENTIFIC FIELD

NUMBER OF STUDENTS AND FELLOWS	FORMAL SCIENTISTS	PHYSICAL SCIENTISTS	MOLECULAR BIOLOGISTS
None	41%	13%	0%
1	15	13	10
2 or 3	33	35	10
4 or more	11	39	80
TOTAL	100	100	100
(Number of cases)	(27)	(23)	(10)

students and fellows is discussed at greater length in Chapter III. Molecular biologists are far more likely than other scientists to have postdoctoral fellows in their research groups. In preparing tables, scientists with two or more students have been contrasted with those who have only one, or no, students.

(7) *Proportion of work done in collaboration with others.* Each informant was asked to estimate what proportion of all his published work had been written with someone else. Although some estimates were uncertain, the sample was divided into three groups: High, 61 per cent or more of all papers; medium, 20 to 60 per cent; and low, less than 20 per cent.

NOTES

1. Since many readers will be scientists, they will share these values and may find it difficult to conceive how the idea of "disinterested curiosity" strikes others. The writings of Friedrich Nietzsche and other antirationalist philosophers are helpful in this regard; Nietzsche viewed disinterested curiosity as a form of

psychopathology. See, for example, the sections, "Immaculate Perception" and "Scholars," in *Thus Spake Zarathustra;* the sections, "Prejudices of Philosophers" and "We Scholars," in *Beyond Good and Evil;* and sections 23 through 25 of his essay, "Ascetic Ideals."

2. Berelson asked recent recipients of the Ph.D. whether they had learned "a great deal from one another" as graduate students. "About three-fourths said they had—a little more in the sciences, a little more in the top places (where the better students are). As a matter of fact, when I went on to ask, 'When you get right down to it, and taking everything into account, did you learn more from your fellow students or from your professors?', only about three-fourths said their professors. Most of the others said the score was about even." Bernard Berelson, *Graduate Education in the United States* (New York: McGraw-Hill, 1960), p. 105.
3. Cf. the following articles by Howard S. Becker and James Carper: "The Development of Identification with an Occupation," *American Journal of Sociology*, 61 (1956), 289–298; and "The Elements of Identification with an Occupation," *American Sociological Review*, 21 (1956), 341–348. They point out that engineers are neither as likely as scientists to develop a distinctive sense of identity nor as isolated in their educational careers as scientists are. This is related to the fact that engineering, unlike medicine or science, is more likely to be perceived as a step to a different occupation than as a permanent one.
4. Thomas S. Kuhn, "The Essential Tension: Tradition and Innovation in Scientific Research," in Calvin W. Taylor and Frank Barron, eds., *Scientific Creativity: Its Recognition and Development* (New York: Wiley, 1963), pp. 341–354, on pp. 344 f. Kuhn goes on to suggest that, incompatible as it is with the values of liberal education, this kind of education seems to work—the sciences that have it develop more rapidly than those that do not.
5. Research during crisis and revolutionary situations is something else again and will be discussed in Chapter VI.
6. Kuhn, *op. cit.*, p. 348; see also Kuhn, *The Structure of Scientific Revolutions* (Chicago: University of Chicago Press, 1962), chs. III and IV.
7. See Berelson, *op. cit.*, pp. 167–171, and James A. Davis, *Stipends and Spouses* (Chicago: University of Chicago Press, 1962), pp. 109–115, 264.
8. This sometimes seems to hold true in other professional contexts. An engineering executive in an electronics firm complained that the engineers under him did not pay adequate attention to the literature. He described them as suffering from the "N.I.H. factor": the belief that if something was "not invented here," in this firm, in this department, or even in this very workroom, it did not exist. As a result, much time was wasted inventing things that had already been invented in the firm or in other firms. Unlike scientists, engineers are not as closely bound into a larger professional community.
9. When this is so, editorial decisions to publish are kept independent of the possibility of payment. Thus, in 1962, only 78 per cent of the pages published in the *Journal of Mathematical Physics* were paid for by the authors' institutions. See Henry A. Barton, "The Publication Charge Plan in Physics Journals," *Physics Today*, 16, 6 (June 1963), 45–57.
10. Neil J. Smelser, "A Comparative View of Exchange Systems," *Economic Development and Cultural Change*, 7 (1959), 173–182.
11. Cf. Alvin W. Gouldner, "The Norm of Reciprocity," *American Sociological*

Review, 25 (1960), 161–178; and Marcel Mauss, *The Gift: Forms and Functions of Exchange in Primitive Societies* (Glencoe, Ill.: Free Press, 1954), pp. 40 f., 73, *et passim.*

12. Mauss, *op. cit.*, p. 21, reporting the work of Malinowski.
13. In a series of interviews with twenty eminent American biologists, Anne Roe was given the same impression about the suppression of the wish for recognition: ". . . the concentration is on the work primarily as an end in itself, not for economic or social ends, or even for professional advancement and recognition, although they are not indifferent to these." Roe, "A Psychological Study of Eminent Biologists," *Psychological Monographs*, 65 (1951), p. 65. Bernice T. Eiduson makes a similar report based on her study of forty scientists in her *Scientists: Their Psychological World* (New York: Basic Books, 1962), pp. 162 and 178 f. See also Charles Darwin, *The Autobiography of Charles Darwin, 1809–1882*, Nora Barlow, ed. (London: Collins, 1958), p. 141, for another scientist's disavowal of desire for recognition.
14. Mauss, *op. cit.*, p. 3 *et passim.*
15. Merton, "Priorities in Scientific Discovery," *American Sociological Review*, 22 (1957), 635–659.
16. Italics in quotations are the interviewer's questions or comments. Statements in brackets are my paraphrases or additional comments.
17. This may be a common failing of eminent men. Leopold Infeld wrote, regarding Einstein; "Before we published our paper I suggested to Einstein that I should look up the literature to quote scientists who had worked on this subject before. Laughing loudly, he said: 'Oh yes. Do it by all means. Already I have sinned too often in this respect.' " *Quest: The Evolution of a Scientist* (New York: Doubleday Doran, 1941), p. 277.
18. See Kuhn, *The Structure of Scientific Revolutions, op. cit.*, pp. 35–42. The layman can get some ideas of the gratification involved by reading such novels as C. P. Snow's *The Search* or Sinclair Lewis' *Arrowsmith.*
19. Sprat, *The History of the Royal Society of London* (London, 1673), pp. 74 f. See also Karl Mannheim on the importance of the desire for recognition in science and other cultural pursuits: *Essays on the Sociology of Knowledge* (London: Routledge and Kegan Paul, 1952), ch. VI, especially pp. 239, 242–243, 272.
20. Eric T. Bell, *The Development of Mathematics* (2nd ed.; New York: McGraw-Hill, 1945), p. 153.
21. Compare the reception by chemists of chromatographic techniques, discovered by Tswett, a Russian botanist, in 1906: ". . . the chromatographic method got off to a bad start . . . for a lowly botanist to assault thus the whole chemical profession was unthinkable! . . . the chromatographic method fell largely into disrepute, and . . . Tswett spent the later part of his life in misery and poverty." The importance of the technique was only recognized in the late nineteen twenties. James E. Meinhard, "Chromatography: a Perspective," *Science*, 110 (1949), 387–392.
22. *Science and the Planned State* (London: George Allen and Unwin, 1945), p. 33.
23. Schrödinger, in *Science and the Human Temperament*, James Murphy, trans. (London: George Allen and Unwin, 1935), pp. 76–80.
24. Cf. the comment of a referee of a biochemical journal: "If the editor reversed what I said he would have to look for a new referee awfully fast, because I don't take that job lightly. And I've turned down many papers, too."

25. See Chapter III with regard to these norms and Chapter IV for a discussion of situations in which exercise of this type of authority leads to conflict.
26. In a survey by the International Council of Scientific Unions of one hundred fifty-six editors of well-known primary journals, 16 per cent reported that manuscripts were not sent to referees and 8 per cent gave equivocal answers to a question on the topic. Reported by J. R. Porter, "Challenges to Editors of Scientific Journals," *Science,* 141 (1963), 1014.
27. Simon Marcson shows how organizations "structure recognition by means of appropriate symbols, including titles, size of office, accessibility, financial rewards, and so on." *The Scientist in Industry* (New York: Harper, 1960), p. 73. However, the difference between recognition and other rewards should be kept clear, for otherwise it is difficult to analyze social control and to specify the source of control. In this work "recognition" means only the written and verbal behavior and the "expressive gestures" of scientists that indicate their approval and esteem of a colleague because of his research accomplishments.
28. Put another way, flexibility and rationality are maximized when workers are alienated from their products. Cf. Talcott Parsons, "*Voting* and the Equilibrium of the American Political System," in Eugene Burdick and Arthur J. Brodbeck. eds., *American Voting Behavior* (Glencoe, Ill.: Free Press, 1959), p. 89. See also Max Weber's stress on "formally free" and actually alienated labor as a defining characteristic of capitalism: *General Economic History* (Glencoe, Ill.: Free Press, 1927), p. 277.
29. Talcott Parsons, *Essays in Sociological Theory* (Glencoe, Ill.: Free Press, 1954), ch. II.
30. This does not mean that scientists are not supposed to be skeptical; that they should be skeptical about their own work as well as that of their colleagues is one of the more important institutionalized norms of science. Cf. Robert K. Merton, *Social Theory and Social Structure* (Glencoe, Ill.: Free Press, 1949), pp. 315 f. It does mean that unlike the consumer in the free market, the "consumer" of scientific products can hold the producer morally responsible for "defective products."
31. Cf. Merton, "Priorities in Scientific Discovery," *op. cit.*, pp. 640 f., and *Social Theory and Social Structure, op. cit.*, pp. 312 f. Mauss pointed out that members of some "archaic societies" felt gifts somehow remained part of the donor and that this belief was reinforced by further existential beliefs, e.g., the gift itself possessed the power to harm the recipient if it was not reciprocated. Mauss, *op. cit.*, pp. 41 ff. *et passim.*
32. Cf. Merton, "Priorities in Scientific Discovery," *op. cit.*, pp. 642–644. See also Edwin G. Boring, "Eponym as Placebo," in his *History, Psychology, and Science: Selected Papers* (New York: Wiley, 1963), pp. 5–28, where it is suggested that eponymy has psychological benefits even for scientists who cannot hope to be remembered this way themselves. Compare Edward A. Gall's questionnaire study of medical scientists, in which he attempted to determine the extent to which they favored or opposed the use of eponymous terms: "The Medical Eponym," *American Scientist,* 48 (1960), 51–57. His results were inconclusive, which is not surprising since incentives of this type are seldom adopted on rational grounds or evaluated according to technical standards.
33. As Georg Simmel says, gratitude "establishes the bond of interaction, of the reciprocity of service and return service, even where they are not guaranteed by external coercion." *The Sociology of Georg Simmel,* Kurt Wolff, ed. and trans. (Glencoe, Ill.: Free Press, 1950), p. 387. He goes on to note that persons make

great efforts to avoid receiving gifts in order not to make such commitments to others. Something like this may occur in science.

34. In other words, the scientific contribution is closely approximated by the sacrificial model of the gift. Cf. Émile Durkheim, *The Elementary Forms of Religious Life*, J. W. Swain, trans. (London: George Allen and Unwin, 1915), pp. 342 f.: "The sacrifice is partially a communion; but it is also, and no less essentially, a gift and an act of renouncement."
35. Quoted in Arthur Koestler, *The Sleepwalkers* (London: Hutchinson, 1959), pp. 393 f.
36. Hans H. Gerth and C. Wright Mills, trans. and eds. *From Max Weber: Essays in Sociology* (New York: Oxford University Press, 1946), pp. 132, 134. Weber was partly concerned with the particular aspects of science in German universities, but the entire essay shows that he was also concerned with the more universal aspects of science as a profession.
37. Cf. Eric T. Bell, *Men of Mathematics* (New York: Simon and Schuster, 1937), for these and others.
38. The story of Job is the classic version of the hero who serves without reward.
39. Serious, unorthodox errors in a text might harm an author's reputation, although certain kinds of orthodox errors may be permitted as "pedagogical license." In any case, although an author's reputation may help sales of a text, it is possible for a person who has no reputation as a research scientist to write a text that will be widely adopted. One of my informants was such an author.
40. See also "Instructions to Appointment and Promotion Committees," *University Bulletin* (California), 10 (1962), 114.
41. Recognition by colleagues is probably a more effective means of social control among scientists, whose products are necessarily public knowledge, than it is among some other professionals. The limitations of colleague control in medicine are pointed out in Eliot Freidson and Buford Rhea, "Processes of Control in a Company of Equals," *Social Problems*, 11 (1963), 119–131.
42. In a survey of scientists (mostly biological and medical) in a large government research organization, 72 per cent of more than three hundred answered "Always" to the question, "In scientific or other professional papers about work to which you have made some contribution, is proper credit given to your own contribution by means of authorship or acknowledgment?" Barney G. Glaser, "Variations in the Importance of Recognition in Scientists' Careers," *Social Problems*, 10 (1963), 273.
43. This man had received more recognition for a different part of his thesis, which he himself felt was not especially important. The difference was that recognition came from a different source: "Another part of my thesis, that dealing with slow neutrons, has been important industrially for reactor physics. I don't know exactly why, but I have a reputation among reactor physicists. I have been invited to work summers at a large establishment as a result of this work." Nonspecialists as a source of recognition are discussed later.
44. Compare the novelist's presentation of a similar case in C. P. Snow's *The Search* (London: Macmillan, 1958).
45. Cf. J. E. Littlewood's proposition (which he attributes to A. S. Besicovitch): "A mathematician's reputation rests on the number of bad proofs he has given." *A Mathematician's Miscellany* (London: Methuen, 1953), p. 41. Pioneer work is often clumsy, and Littlewood gives some vivid examples of this.
46. A directory of such awards for the United States has been edited by Margaret A.

Firth, *Handbook of Scientific and Technical Awards in the United States, 1900–1952* (New York: Special Libraries Association, 1956).

47. Flora Kaplan, *Nobel Prize Winners* (Chicago: Nobelle, 1941).
48. Paul F. Lazarsfeld and Wagner Thielens, Jr., have shown that, in the social sciences, the correlation between productivity and having held office in professional societies is high but not perfect—many (21 per cent) of those with the highest productivity scores have not held office, and many (23 per cent) of those with the lowest scores have. *The Academic Mind* (Glencoe, Ill.: Free Press, 1958), p. 9. Similar findings, for a sample of biologists, political scientists, and psychologists, are presented in Diana M. Crane, "The Environment of Discovery" (Doctoral dissertation, Columbia University, 1964), pp. 74 f.
49. The Gordon conferences cover a wide range of topics in the natural sciences and technology. They were initiated by Prof. Neil Gordon in 1931 and are limited to at most 100 scientists per conference. (Even so, with at least thirty-six conferences every summer, many scientists attend.) Participation is by invitation only, and eminent men usually act as conference chairmen. The style of the conferences is highly informal; each one lasts a week, during which there are only eighteen hours of "formal" discussion. Informality is assured by the practice of keeping the conferences "off the record"; proceedings are never published, what participants say is not quoted, and newspapermen are barred from the small New England colleges where the conferences are held. See "Doctor Gordon's Serious Thinkers," *Saturday Review of Literature*, 39 (August 4, 1956), 42–46.
50. For a general discussion and a tentative statement of the communications channels most apt to be used for conveying different forms of information, see Herbert Menzel, *The Flow of Information among Scientists* (New York: Columbia University Bureau of Applied Social Research, 1958). This research is summarized in Menzel, "Planned and Unplanned Scientific Communication," in Bernard Barber and Walter Hirsch, eds., *The Sociology of Science* (New York: The Free Press of Glencoe, 1962), pp. 417–441. A review of technical studies of scientific communication is given by Helen L. Brownson, "Research on Handling Scientific Information," *Science*, 132 (1960), 1922–1931.
51. For example, in most mathematical journals the waiting time between receipt of the final revision of a manuscript and its publication is a year or more, because of the large backlog of articles. See the survey reported in the American Mathematical Society *Notices*, 8 (February 1961), 42.
52. "To accept [a gift] without returning or repaying more is to face subordination, to become a client and subservient, to become *minister*." Mauss, *op. cit.*, p. 73. See also Peter Blau's study of a federal enforcement agency, *Dynamics of Bureaucracy* (Chicago: University of Chicago Press, 1955), and subsequent analyses of Blau's results by George C. Homans, *Social Behavior: Its Elementary Forms* (New York: Harcourt Brace and World, 1961). In the office studied, agents were expected to work independently; when problems arose, the agent was expected to consult his supervisor. Agents hesitated to do this, however, and were more likely to consult their formal peers. The more competent agents were asked for such advice on many occasions, and those who complied had the highest informal status in the group. This status was expressed in compliments from inferiors and other forms of behavior in which the agent asking for advice expressed his recognition of the superiority of the other. Furthermore, agents with high status could seek the assistance of others without having to confess inferiority, and their judgments with respect to some agency policies and some informal activities tended to be accepted by the others.

53. This makes it relatively easy to interview scientists for sociological purposes. Very few individuals refused to see me, and most were very generous with their time. In this situation, the interviewer was defined as a scientist seeking information about which the informant was exceptionally well informed. Roe indicates that she, too, received excellent co-operation from the scientists she asked to see: Ann Roe, "A Psychological Study of Eminent Biologists," *op. cit.*; "A Psychological Study of Eminent Physical Scientists," *Genetic Psychological Monographs*, 43 (1951), 121–239; and "A Psychological Study of Eminent Psychologists and Anthropologists and a Comparison with Biological and Physical Scientists," *Psychological Monographs*, 67, 2 (1953). It may be noted that few scientists, unlike members of other elites, have secretarial personnel to turn away those asking for favors.
54. Peter Axel, "Scientific Exchange Visits," *Bulletin of the Atomic Scientists*, 16 (1960), 212–215. The purpose of this article was to urge that Russian scientists be permitted to visit the United States even if reciprocal permission were not granted.
55. *In and Out of the Ivory Tower* (Seattle: University of Washington Press, 1960), pp. 233 f.
56. *I Am a Mathematician* (Garden City, N.Y.: Doubleday, 1956), p. 329.
57. Alvin M. Weinberg, "Impact of Large-Scale Science on the United States," *Science*, 134 (1961), 161–164.
58. "Editorial," *Physical Review Letters*, 4 (January 1, 1960), 2. The American Physical Society has a full-time public relations director. One function of such a position is probably to give all physicists equal opportunities for the esteem of the lay public and to inhibit competition for such esteem.
59. Cf. Barney G. Glaser, *Organizational Scientists: Their Professional Careers* (Indianapolis, Ind.: Bobbs-Merrill, 1964), pp. 7–10.
60. Cf. William Kornhauser, *Scientists in Industry: Conflict and Accommodation* (Berkeley: University of California Press, 1962), ch. V.
61. See Alvin W. Gouldner, "Cosmopolitans and Locals: Toward an Analysis of Latent Social Roles," *Administrative Science Quarterly*, 2 (1957–1958), 281–306, 444–480.
62. See Berelson, *op. cit.*, p. 115. For a summary of research from the nineteen thirties and a general discussion, see Walter A. Lunden, *The Dynamics of Higher Education* (Pittsburgh: Pittsburgh Printing Co., 1939), pp. 316–323.
63. In a study of a government organization conducting basic research in the medical sciences, Glaser found that the scientists strongly committed to the organization also tended to be strongly committed to the goals of basic science. He concluded, "these scientists are both cosmopolitan and local oriented." Barney G. Glaser, "The Local-Cosmopolitan Scientist," *American Journal of Sociology*, 69 (1963), 256. See also Glaser, *Organizational Scientists: Their Professional Careers*, *op. cit.*, ch. 2.
64. Theodore Caplow and Reece McGee, *The Academic Marketplace* (New York: Basic Books, 1958), p. 159.
65. *Ibid.*, p. 107.
66. See also Francis Bello, "The World's Greatest Industrial Laboratory," *Fortune*, 58 (November 1958), 214. Bello shows that only 10 per cent of the papers in the *Physical Review* for 1956 and 1957 came from industry, with only three industrial laboratories contributing a majority of them. (In 1948, 20 per cent of U. S. physicists were employed by private industry, and the proportion had probably in-

creased by 1956. U. S. Bureau of Labor Statistics, *Occupational Mobility of Scientists* [Bulletin No. 1121, Washington: Government Printing Office, 1953], p. 8.) See also Berelson, *op. cit.*, p. 83, for a similar study with similar results.

67. Berelson also showed that scientists in leading universities accounted for a disproportionate number of published papers. He analyzed the authorship of articles in major journals (in the humanities and social sciences, as well as the sciences) for 1958 and reported that the "top twelve universities" are "clearly top on this score of contributing to knowledge. They have less than 10 per cent of the total faculty, but account for almost 40 per cent of the authors in the leading learned journals." Berelson, *op. cit.*, p. 127. Similar results, for social scientists only, were obtained in a nationwide sample by Lazarsfeld and Thielens, *op. cit.*, p. 30.

68. Francis Bello, "The Young Scientists," *Fortune*, 49 (June 1954), 142–182. For another survey of scientists' opinions on this topic, see U. S. President's Scientific Research Board, *Administration for Research* (["Science and Public Policy," I] Washington: Government Printing Office, 1947), pp. 231–232. A still earlier study (1937) found that scientists starred in the *American Men of Science*—an indicator of eminence used in early editions—were disproportionately in universities and were greatly underrepresented in industry and government. U. S. National Resources Committee, Science Committee, *Relation of the Federal Government to Research* (["Research—A National Resource," I] Washington: Government Printing Office, 1958), p. 170.

69. Charles V. Kidd points out that "the federal agencies have put money where research capacity has been demonstrated." Of federal research funds supplied to universities, fourteen universities received 55 per cent, and twenty-five universities received 66 per cent of the total in 1952. *American Universities and Federal Research* (Cambridge: Harvard University Press, 1959), pp. 54–60. He goes on to note that the decisions to allocate research funds are frequently made by university scientists. The important decisions tend to be made through the advisory groups of federal grant-giving agencies. Most members of such groups—63 per cent—are from universities (p. 199). Because of such groups, Kidd concludes, "Never have the scientists of the nation had so much authority over the funds which they receive" (p. 201).

70. See Berelson, *op. cit.*, pp. 113–115, which indicates that faculty of high-quality institutions tend to be recruited from the graduates of such institutions.

71. *Ibid.*, pp. 111 f.

72. For data on the inferior salaries of physicists and chemists in universities as compared with those in industry, see U. S. National Science Foundation, *American Science Manpower 1954–1955* (Washington: Government Printing Office, 1959), p. 10. Such income differences certainly should not be exaggerated; the income of American university professors is often very high in absolute terms. Cf. Seymour M. Lipset, *Political Man* (Garden City: Doubleday, 1960), pp. 334 f. A survey of basic and applied agricultural scientists in a state experiment station showed that the basic scientists were far less likely to place great importance on salary in evaluating a job than were applied scientists. See Norman W. Storer, "Research Orientations and Attitudes toward Teamwork," *IRE* [Institute of Radio Engineers] *Transactions of the Professional Group on Engineering Management*, EM-9 (1962), 29–33.

73. For data on this, see Kornhauser, *op. cit.*, pp. 71 f., *et passim;* and Marcson, *op. cit.*, pp. 52–57.

74. Joseph Ben-David, "Scientific Productivity and Academic Organization," *American Sociological Review*, 25 (1960), 828–843.

75. *Ibid.*, p. 840. See also Thorstein Veblen, *The Higher Learning in America* (New York: Sagamore Press, 1918), pp. 77, 82–85, *et passim,* on the consequences of rivalry between universities. Veblen takes a far less sanguine view of these consequences than Ben-David does.
76. See, e.g., Caplow and McGee, *op. cit.*, p. 221 *et passim.*
77. An assistant professor of mathematics in a western university had a lengthy list of publications, but since most or all of them had been published in unrefereed journals, or had been the result of collaboration with more eminent men, he did not receive a permanent position. There was a possibility of his being promoted because he was a very popular teacher, but if it did occur it would be over the strong protests of other mathematicians in the department.
78. The role-set is defined as "that complement of role relationships which persons have by virtue of occupying a particular social status." Robert K. Merton, *Social Theory and Social Structure* (rev. ed.; Glencoe, Ill.: Free Press, 1957), p. 369. See also his general discussion, *ibid.*, pp. 368–389.
79. U.S. Bureau of Labor Statistics, *Employment, Education and Earnings of American Men of Science, op. cit.* See also Anselm Strauss *et. al.*, *The Professional Scientist* (Chicago: Aldine, 1962), pp. 234 f.
80. Veblen, *op. cit.*, p. 200. A study of 552 university, government, and industrial scientists tends to confirm this. Productivity among scientists in the sample was higher when they spent 41 to 80 per cent of their time on research than when they spent more than 80 per cent. However, scientists who spent the remainder of their time on administrative work were more productive than those who spent the remainder of their time teaching. Frank M. Andrews, "Scientific Performance as Related to Time Spent on Technical Work, Teaching, or Administration," *Administrative Science Quarterly*, 9 (1964), 182–193.
81. For contrary views, see the following: J. D. Bernal, *The Social Function of Science* (London: Routledge, 1939), p. 105; Charles Babbage, *Reflections on the Decline of Science in England and Some of Its Causes* (London: B. Fellowes, 1830), pp. 19 f.; and Sprat, *op. cit.*, pp. 67 ff. It is often felt that research actually benefits teaching. The graduate faculties in U. S. universities evidently think so; see Berelson, *op. cit.*, pp. 152–157, 289, and 292. Thorstein Veblen (*op. cit.*, pp. 12 f.) strongly supported this position. Moreover, it is by no means clear that lack of emphasis on research benefits teaching. See, for example, Adam Smith's comments on the teaching in eighteenth-century English universities, which were hardly research-oriented institutions, in *Wealth of Nations* (Edinburgh: Black, 1863), pp. 343–350.
82. That is, such persons will probably serve as the primary "reference group" for the scientist. See Robert K. Merton, *Social Theory and Social Structure, op. cit.*, pp. 225–386; and Tamotsu Shibutani, "Reference Groups as Perspectives," *American Journal of Sociology*, 60 (1955), 562–569.
83. Emilio Segrè, "Biographical Introduction," *Enrico Fermi: Collected Papers* (Chicago: University of Chicago Press, 1962), pp. xxxi–xxxii, xli.
84. Cf. Gouldner, "Cosmopolitans and Locals," *loc. cit.*
85. See, for example, Bernal, *op. cit.*, pp. 113–116.
86. A survey supporting this point is reported in Lazarsfeld and Thielens, *op. cit.*, p. 9.
87. Infeld, *op. cit.*, p. 106.
88. Wiener, *op. cit.*, p. 360.

89. See the biography given by Hardy in James R. Newman, ed., *The World of Mathematics* (New York: Simon and Schuster, 1956), pp. 368–380; also the discussions in Hardy, *A Mathematician's Apology* (Cambridge: Cambridge University Press, 1941); and Littlewood, *op. cit.*
90. Infeld, *op. cit.*, pp. 274 f.
91. *Ibid.*, p. 285.
92. It is also interesting to note that about one-third of the papers his bibliographer classifies as his "principal works" were written jointly—although most of these appeared in his later life, when in fact his productivity had declined. Computed from the Bibliography in Paul A. Schilpp, ed., *Albert Einstein: Philosopher-Scientist* (Evanston, Ill.: The Library of Living Philosophers, 1949). The situation of Isaac Newton, another relatively isolated theorist, is also worth noting. As with Ramanujan, social influences may not have played a decisive role in the formation of his ideas, although he was a student and protégé of one of the leading English mathematicians of the time, Barrow. Even so, as with Ramanujan, the informal influence of others, notably Halley, was crucial in bringing his work to the attention of others and in bringing him the recognition he deserved. For a psychoanalytic interpretation of the sources of Newton's tendency to isolate himself, see Lewis S. Feuer, *The Scientific Intellectual* (New York: Basic Books, 1963), pp. 411–419.
93. See the appendix to this chapter for details on measurement procedures.
94. Diana Crane, *op. cit.*, p. 137, reports that, in her sample, "Talking to departmental colleagues was not related to productivity."
95. "Some Social Factors Related to Performance in a Research Organization," *Administrative Science Quarterly*, 1 (1956), 310–325.
96. *Ibid.*, p. 315.
97. On the measurement of productivity and its association with recognition, see D. C. Pelz, "Relationships between Measures of Scientific Performance and Other Variables," in Taylor and Barron, *op. cit.*, pp. 302–310; and Glaser, *Organizational Scientists: Their Professional Careers*, *op. cit.*, pp. 19–23.
98. Some evidence that the same factors affect scientific productivity in universities and government laboratories is presented in Andrews, *op. cit.*
99. The general outlines of this thesis have also been expressed by Michael Polanyi. He writes, for example: "The opinion of the [scientific] community exercises a profound influence on the course of every individual investigation. Broadly speaking, while the choice of subjects and the actual conduct of research is entirely the responsibility of the individual scientist, the recognition of claims to discoveries is under the jurisdiction of scientific opinion expressed by scientists as a body. Scientific opinion exercises its power largely informally. . . ." *The Logic of Liberty* (London: Routledge and Kegan Paul, 1951), pp. 34 f. Elsewhere, Polanyi compares the organization of science with free-market organizations, perhaps in an attempt to convince us of the virtues of the latter: "Pure and Applied Science and Their Appropriate Forms of Organization," in Congress for Cultural Freedom, *Science and Freedom* (London: Martin Secker and Warburg, 1955), pp. 39–49. He does not consider the possibility that the pathologies of the capitalist free market have their analogous forms in science; for an elaboration of such possibilities, see the following chapters.
100. This would be true only for scientists without established reputations. Distinguished scientists may be able to change specialties and carry their prestige with them. Similarly, scientists may be able to move from a discipline of high pres-

tige to one with less prestige and carry some of their original prestige with them. See the discussion in Chapter IV.

101. In U. S. graduate schools, productivity does not decline with rank, as the contractual theory would seem to imply. This is indicated by data from Berelson's study of graduate education. (Table not shown.)

102. Caplow and McGee, *op. cit.*, p. 221.

103. Eric Ashby, "Universities Today and Tomorrow," *Listener, 65* (June 1, 1961), 959.

104. Berelson has already done this well, *op. cit.*

105. This has been more or less proposed by J. D. Bernal, *op. cit.*, and by other English Marxists. The point of view has been strongly objected to by Michael Polanyi (see note 99 for this chapter) and J. R. Baker, *op. cit.*, mostly on a priori bases. Bernal and others have presented very rosy descriptions of Soviet science as planned science. Other descriptions, suggesting that Soviet basic science, in so far as it is effective, is relatively unplanned and informally organized, can be found in Eric Ashby, *Scientist in Russia* (Harmondsworth Middlesex: Pelican, 1947); David Joravsky, *Soviet Marxism and Natural Science 1917–1932* (New York: Columbia University Press, 1961); and Alexander Vucinich, *The Soviet Academy of Sciences* (Stanford: Stanford University Press, 1956).

106. Cf. Kornhauser, *op. cit.*, pp. 32–41, 56–80.

107. For a more complete discussion of the problem of measuring scientific productivity, see Crane, *op. cit.*, pp. 38–44.

II

COMPETITION FOR RECOGNITION[1]

Recognition is normally given for the first formal presentation of an innovation or discovery to the scientific community. Some recognition may also be awarded for replications and rediscoveries, although their value to scientists is less than the value of original work and recognition is correspondingly less. The system of incentives in science does not encourage workers to devote their efforts to repeating past accomplishments when the record of such accomplishments is available in libraries. Professional scientists whose research is likely to lead to the same discoveries are therefore invariably either competitors or collaborators.

Three definitions are implicit in the preceding generalization. A "professional scientist" is one to whom discoveries are attributed and to whom the recognition for them is awarded. He is the individual who selects the research problem and the methods to be used and decides whether to accept or reject the initial results. Recognition for specific discoveries is awarded not so much for the skills required for making the discovery, since some discoveries considered important are technically "easy," or for the effort put into the research, since the scientist may delegate most of the dirty work to students and technicians. Rather, recognition is awarded to the individual who exercises the decision-making functions, who freely selects problems and methods, and who evaluates the results. Although there is no a priori reason for having a single individual select the problem and methods and decide whether to accept the results, this is the typical procedure in basic research. It is less prevalent in applied research and in some types of modern basic research.

Competition results when two or more scientists or groups of scientists seek the same scarce reward—priority of discovery and the recognition awarded for it—when only one of them can obtain it. Competitors need not be aware of one another's existence. Collaboration occurs when two or more individuals consciously co-operate in seeking a scarce reward and share it, if and when it is obtained. In science this usually means joint authorship of publications announcing discoveries and shared recognition for them. Collaboration will be considered in the next chapter. Here I shall discuss competition in science, its prevalence, severity, and consequences.

The Prevalence and Severity of Competition

Competition is manifested, to the working scientist as well as to the observer of science, in the experience of being anticipated in the presentation of research results, or, in the vernacular, in being "scooped." This occurs when one scientist has selected a problem, begun research on it, and, perhaps, essentially solved it and another scientist publishes a solution before the former is able to do so. (I also include as anticipation those cases in which a scientist works for some time on a problem, only to find that a solution has already been published. The subjective aspect of anticipation is important here.) Competition varies in its prevalence and severity. When scientists are frequently anticipated and can expect to be anticipated often, competition is prevalent; when they are never anticipated, competition is nonexistent. Severity refers to the consequences of anticipation: if being anticipated results in a scientist's inability to publish his results and thereby gain recognition for his work, competition will be severe; if the publication of anticipated work is not only permitted but actively sought, competition is mild. The prevalence of competition is not necessarily related to its severity, except that competition cannot be severe if it is nonexistent.

Another manifestation of competition is simultaneous discovery. Naturally, this occurs far less frequently than anticipation. Nevertheless, it had been experienced in a clear form by three of the sixty-six scientists to whom I spoke on this topic. The report of a mathematician illustrates the dramatic nature of such events:

> I gave an invited lecture once at Stanford, at a meeting of the national organization in that part of the country, and there was another meeting on the same day in New York, at which another man was speaking. While this other fellow and I had had no contacts with each other, it turned out that when we turned our papers in there was quite a big overlap. This was a little upsetting at first, because I had laid quite a store by that, but I became very good friends with this fellow.[2]

Simultaneous discovery is unlikely to have severe effects; both contributors usually receive credit for the work. For example, a mechanism in nuclear magnetic resonance is named after both a theoretical physicist who was interviewed and its simultaneous discoverer, in the form "the Smith-Jones mechanism."[3]

The experience of simultaneous discovery is relatively rare, and it will not be considered further here. Anticipation is far more frequent.[4] This had been experienced at least once by forty of the scientists interviewed, or by 61 per cent of the sixty-six scientists for whom information on this point was obtained.[5] The question, "Are you concerned about the possibility of being anticipated in presenting the results of your current research?" was asked of sixty-two scientists, and 50 per cent did express concern. Those who had experienced anticipation were somewhat more likely to be concerned about its occurring in the future, but the correlation was low (Q= .24, N = 61). Fewer scientists, then, were concerned about being anticipated than had actually had the experience, and many who could *expect* to be anticipated were not *concerned* about it. This is so because the consequences of anticipation are often mild and because it may be thought improper to be concerned about the problem. Before discussing variations in the severity of consequences, let us briefly consider other variables that affect concern and anticipation.

Obviously, the longer a scientist has been conducting research, the greater his likelihood of having been anticipated. Because of confounding factors, no doubt, the association between age and experience of anticipation was low in the sample of scientists interviewed for this study. If the sample is divided into those who received the Ph.D. before 1946 and those receiving it in 1946 or later, the association between age and experience of anticipation is only Q = .09 (N = 65).[6] However, concern over anticipation is greater for the younger men, as shown in Table 8.[7] Older men are more likely to have an established reputation, and being anticipated on a

TABLE 8

DATE OF DOCTORATE AND CONCERN ABOUT ANTICIPATION*

"Are you concerned about the possibility of being anticipated?"	DATE OF RECEIVING Ph.D.		
	Before 1946	*1946 or Later*	*Total*
Yes	29%	65%	50%
No	71	35	50
TOTAL	100	(37)	(62)
(N)	(24)	100	100

* Q = –.64.

single piece of work is not likely to affect it very much, unless the discovery is of the greatest importance. Younger men are likely to be seeking a reputation and a position, and being anticipated may negate the value of more than a year's work—a serious setback to them. However, there is probably more to it than this. The concern of the younger man is often unrealistic. For example, a young theoretical physicist, in response to the question about his concern over being anticipated said: "I was as a thesis student, but such fears were purely imaginary, and now I am not concerned. There is lots of communication and I am not too worried about some unknown doing it unless he's in Japan or the Soviet Union." A number of other scientists reported the same attenuation in concern as their careers progressed.[8] An elderly theoretical physicist said: "When you think of a good idea, you immediately feel that at least twenty others will have the same idea. But as I grow older I find that this isn't so; in my experience it is rather rare that this happens."

Scientists are concerned about being anticipated because they hope the solutions to their problems will interest other scientists, and, given this hope, it is only reasonable to expect other scientists to approach the same problems. As time goes on, however, the scientist learns of ways in which the threat of being anticipated can be reduced. He also learns that being anticipated does not always or necessarily destroy the value of his work.

The concern over being anticipated arises because the scientist is consciously oriented toward achieving recognition and is afraid of not receiving it for work he has actually done. But in the preceding chapter it was noted that the conscious pursuit of recognition was considered inappropriate for scientists; it is felt that their discoveries should be given freely and magnanimously to the community. As a result, many scientists disparage competitive behavior and will not condone it even if they are in competitive situations themselves. Thorstein Veblen, an academic nonconformist and opponent of the competitive ethos of American life, put it this way:

> By tradition the faculty is the keeper of the academic interests of the university and makes up a body of loosely-bound non-competitive co-partners, with no view to strategic team-play and no collective ulterior ambition, least of all with a view to engrossing the trade. . . . Learning is, in the nature of things, not a competitive business. . . .[9]

More often, those who have achieved high rewards in competitive situations will be in a better position to condemn competitive behavior. In this connection, a Nobel-Prize-winning physicist said:

> In my opinion, if it is only a race it is not worth while doing. If it is only a matter of posting a letter today as against posting one tomorrow, it is not important. If one goes into physics with the idea of running races, he

> is more likely to do something stupid. Many wrong papers are published because of this. It sometimes borders on bad faith—people will publish before they are sure of results.

Sometimes the wish is father to the thought. Another Nobel-Prize-winner, who had been anticipated in presenting one of the great discoveries of modern physics, asserted, "I don't think there is any such thing as competition in physics." In fact, however, the characteristics of physical research make it the most competitive scientific field. Let us now consider the characteristics of scientific disciplines that influence the prevalence and severity of competition within them.

DISCIPLINARY DIFFERENCES IN THE INTENSITY OF COMPETITION

Competition results when scientists can agree on the relative importance of scientific problems and when many of them are able to solve these problems. We can deduce from this that the prevalence and severity of competition will be greater (1) as agreement about the relative importance of problems increases, and (2) as the number of specialists able to attack any given problem increases. A third factor, which determines not the prevalence but the severity of competition, is the degree of precision that can be obtained—the relative degree of confidence specialists may have in particular results. When this degree of confidence is high, replications will be of little value; when it is low, replications may be necessary. Thus, another generalization can be made: (3) the severity of competition will be greater as the degree of confidence in particular research results increases.

On the basis of general information about different disciplines, the relative degree of competition within them can be predicted. Because physics has a highly developed and logically tight theory, as well as highly precise experimental techniques to test its adequacy, physicists can select significant problems with relative ease by noting areas in which theory has not been tested and in which experiment has failed to confirm theory.[10] Furthermore, since the experiments are so precise, relatively great confidence can be given to the first results reported and less recognition will be accorded for replication. Physics is also a well-established discipline, and many physicists are available for work on important problems. (Significant exceptions will be noted.) As a consequence, we can expect to find the greatest prevalence and intensity of competition in physics. Chemistry and molecular biology, on the other hand, have less well-developed theories, and the interpretation of experimental results must often proceed in a highly tentative manner. Nevertheless, there are growing points in these fields—points at which interest tends to be focused; there are many well-trained specialists; and equipment is usually easy to get. Consequently, an intermediate level of competition can be expected in these fields.

Theories are, of course, rigorous and well developed in the formal sciences—mathematics, statistics, and logic; these fields define the standards of logical rigor for the other sciences. In addition, each article presenting a discovery can present almost the entire proof for it, so that each reader may "replicate" the proof for himself; therefore, additional articles proving the same theorem will receive no recognition, with a few exceptions which will be noted. For these reasons, competition in the formal sciences might be expected to be prevalent and severe. However, other characteristics of these fields mitigate the prevalence, if not the severity, of competition within them. The mathematician has great freedom, since experience cannot contradict his results. An infinite number of mathematical systems are possible and can be constructed, and the criteria for calling some of the systems more important than others, and hence some problems more important than others, are vague and not compelling. This is especially so when mathematicians spurn the criteria of practical utility, as they now do in the United States. It is often relatively difficult, therefore, for mathematicians to order problem areas in their field either according to relative importance or according to the recognition that should be awarded discoveries in them. For this reason, the prevalence of competition in mathematics can be expected to be relatively low. The same is true of symbolic logic.

Statistics is closer to experience, and presumably some statistical results could be disproved by experience. (This is, of course, not true of the mathematical theory of probability.) However, statistics, and logic as well, possess two other characteristics that reduce the prevalence and intensity of competition within them. Both are new fields, and both have relatively few practitioners. The first periodical devoted strictly to mathematical statistics was *Biometrika*, founded in 1901, and the first periodical devoted to symbolic logic was the *Journal of Symbolic Logic*, founded in 1936. In both fields many problems of great and fundamental importance remain to be solved. Furthermore, both fields have relatively few practitioners. Mathematical logic, since it is not established as an independent discipline, is in a dependent position in departments of mathematics and philosophy. Statistics, although an independent discipline, has few practitioners. Not only is there a critical shortage of statisticians to meet the needs of the larger society, but, within the field itself, many problems of great intrinsic significance have not yet been approached. In such fields as these, it is easy for scientists to avoid competitive situations.

The following hypotheses seem justified in the light of the foregoing analysis: (1) The prevalence of competition will be greatest in physics, intermediate in chemistry and molecular biology, and relatively low in the formal sciences. (2) The severity of competition will be greatest in the formal sciences, less in physics, and least in molecular biology.[11] These hypotheses are supported, with one minor exception, by a statistical analysis of the interviews of this study. (See Table 9.) The proportion of

TABLE 9

EXPERIENCE OF ANTICIPATION BY SCIENTIFIC FIELD

AREA	PERCENTAGE HAVING BEEN ANTICIPATED	(NUMBER OF CASES)
Physical scientists	70	(27)
Molecular biologists	64	(11)
Formal scientists	48	(25)

scientists who have been anticipated is a direct indicator of the prevalence of competition; this is highest for physical scientists (theoretical and experimental physicists and physical chemists), next highest for molecular biologists, and lowest for formal scientists (mathematicians, statisticians, and logicians). A crude indicator of the severity of competition is provided by the percentage of scientists expressing concern about the possibility of being anticipated in their current research. (Clearly the concern is a function of both the prevalence and the severity of competition, but we may assume provisionally that severity is the more important component.) The formal scientists were more likely to be concerned about the possibility of being anticipated than were the physical scientists, but, contrary to hypothesis, the molecular biologists were also more likely to express concern than the physical scientists. (See Table 10.) The small

TABLE 10

CONCERN ABOUT ANTICIPATION BY SCIENTIFIC FIELD

AREA	PERCENTAGE EXPRESSING CONCERN	(NUMBER OF CASES)
Formal scientists	61	(23)
Molecular biologists	55	(11)
Physical scientists	40	(25)

number of cases, especially of molecular biologists, and the fact that the indicator reflects both prevalence and severity, mean that these results are somewhat tentative. But the general line of reasoning is supported. A larger proportion of physical scientists reports the experience of anticipation than reports being concerned about the possibility, while the reverse is true for formal scientists.

Being anticipated is more likely to forestall publication (and the receipt of recognition) by formal scientists than by experimental scientists. When the experimental work involves determining physical constants, demonstrating new effects, and so forth, replications tend to be desirable and are publishable. Thus, a solid-state scientist said:

> Being anticipated doesn't matter too much. Unless two people agree on a measurement it isn't worth very much anyway. So, if your work agrees with something done earlier, it's a nice job of confirmation; and if it disagrees, that's interesting too. . . .

This is less likely to hold true for mathematical theorems. In any case, differences within disciplines are large, and, as we have seen, other factors affect the rates of being anticipated and of being concerned about it. The following examples from the interview data from different fields will give an impression of typical experiences within particular fields and of the range of experiences within them.

In the last twenty years, experimental investigations into the physics of elementary particles have come to require, in most cases, enormously expensive and scarce research tools, especially high-energy particle accelerators. Before this, although the necessary expenses were not insignificant, the primary factor limiting the number of physicists entering the field was the possession of experimental skills and abilities. Thus, races to achieve results in this area in the twenties and thirties were often close. An English expert estimated that Lord Rutherford, the English physicist and discoverer of the electron shell atom, usually anticipated his rivals by spans of time ranging from a few months to three or four years.[12] After 1940, the scarcity of equipment was so great that some workers more or less monopolized the opportunity to make leading discoveries. For example, one of the physicists interviewed had been at the University of California Radiation Laboratory before taking his present position. He pointed out that he could not have been anticipated while he was there:

> . . . the only machines that could do the experiments were in Berkeley at the time. Before the machine was ever completed, physicists knew which experiments should be performed first.

The same form of monopolistic position exists in other branches of science. There is only one two-hundred-inch telescope, and astronomical discoveries that require such an instrument can be made only by those having access to it. Similarly, a theoretical physicist was doing meteorological research with a large computer. He was not at all concerned about the possibility of being anticipated by others because, he said:

It's simply mechanical. The [other individuals working on similar problems] still await computers and won't get them for another year or two.

In the course of time many particle accelerators have been constructed for "low-energy" nuclear physics, i.e., the energy ranges lower than those necessary for the production of pions. Hence, competition in this area is more prevalent, despite the fact that the problems not yet solved are regarded as among the less important. One physicist working in this area pointed out that he had been anticipated:

It doesn't bother me unless it discloses an error or sloppiness on my part. My philosophy is that good work deserves publication. You do your best, then publish. I may say that our work has stood the test of time, it has not been shown to be wrong.

Similarly, three students who were interviewed were preparing dissertations in particle physics at medium-energy levels, and two of them had already been partially anticipated. In neither case would this forestall publication, both because of the desirability of replication and because of slight differences in experimental procedures.

Solid-state physics is similar to low-energy physics in the prevalence of competition. There are many workers, significant problems are often evident to all, and equipment is easily available. One physicist working in this area said:

Physicists have used these [techniques for investigating the solid state] a great deal; the area is of general interest. . . . I was anticipated about three or four times in my career. . . . Competition is rather high. The apparatus is easy to build and is in fact commercially available. Any Joe Blow can get into it. That means we must exert ourselves more.

Still other branches of physics attract little attention from the larger community of physicists, although some engage in research in these areas. This is true, for example, of parts of optics and acoustics. Because of the low level of interest in these branches, physicists working in them expressed little concern about the possibility of being anticipated and were not likely to have experienced anticipation.

In theoretical physics the problem of the availability of research facilities rarely arises. (Large computers are seldom necessary for such work.) As a result, competition is usually more prevalent; it is also more severe, since replications are unnecessary. When asked if he was concerned about the possibility of being anticipated, one theoretical physicist answered:

I guess I am sometimes, but I generally take the point of view that if I am anticipated the particular idea must have been so close to the surface that there is not much distinction in getting it. Much more annoying is that, once an original idea is put forward, swarms of people take it up and develop it. You don't have a honeymoon period in which to look into the consequences of what is developing; it becomes a rat race, but I guess it advances the state of the art.

This was a frequent theme of scientists in all disciplines: the pressures of competition made it difficult for a single individual to put forth a unified, elaborate, and reasonably complete presentation of a discovery. If one attempted to withhold publication until he had obtained many of the consequences of his discovery, he was in danger of being anticipated in presenting the central idea, whereas, if he presented the central idea, he would be forced to share recognition for its elaboration with others.

The same theoretical physicist also put forward another almost universal theme:

> In my position [competition] is annoying mainly for the pressure it puts on those working with me. I myself don't publish such developments on more basic ideas. But it is annoying for students. This happened in a very dramatic way two years ago. I was working with X, and we were pushing ahead ideas which were basically his. As soon as that happened, I got my students working on them. The best student got his work published first, but the others were all anticipated. This is O.K. for science but inconvenient for the student. It raises the problem about suggesting thesis problems close to the frontier of knowledge. Many physicists think thesis problems should not be close to the frontier for this reason, but I don't like to assign problems of the other kind. I guess this is selfish—I would rather work with a student on something which is interesting to me.

Some mathematicians who were interviewed said they disliked directing the theses of graduate students for this very reason; it was too hard for them to find problems which would not be trivial but which the student could solve before being anticipated.

Thus, in physics it is usually possible to determine which problems are most important—on the "frontiers" of advancing knowledge—and greatest recognition is given for solutions to them. Competition would always be most intense in research on such problems were it not for the fact that some workers have nearly a monopoly on the research facilities necessary for such work. (In most cases, "oligopoly" would be a better word—there are now more than one, albeit few, particle accelerators at the highest energy ranges.[13]) The possibility of a monopoly of facilities does not arise for theoretical physicists, however, and competition is therefore more prevalent.

The relatively undeveloped state of theory in molecular biology makes it harder for workers in this area to identify problems of central importance, and, therefore, there is less competition with respect to such problems. This interpretation of the situation was given by a number of informants. For example, one gave the following response to a question about his concern over anticipation.

> There is very little of that in a field like ours, because the choice of problems is so large in relation to the amount of work done. Also, there is much less continuity and integration in the development of the field. Even in the developed fields of biology, not many problems turn up that people want to compete on. It is beginning to occur in molecular biology. It is a strong feature of biochemistry. It occurs when important results can be expected. People want to get there first when they know what's there. This is rare in biology. *Have you ever had such an experience?* No. I'm not representative. I tend to work on things far out. *Why?* I enjoy the exploratory phases of a problem, not the exploitation. I keep moving rather than cultivating the ground I plow.

This scientist also expressed a common reaction to competition, that of moving into different areas. This practice, sometimes called "skimming the cream" off a newly developing area, is also common in other such fields as physics and mathematics.

As this informant pointed out, in some areas of molecular biology, especially biochemistry, competition is more prevalent. However, it is not necessarily severe. For example, when the suggestion was made to a biochemist that competitive pressures encourage hasty publication, he responded:

> My students have sometimes said that to me, and I've said, "The world is wide, there are lots of problems. No two people ever do *exactly* the same thing—well, not never, but rarely—and even if someone else publishes the same thing, there is still room for yours." You may have to change the focus a little, but we never hurry into print. *You sound as if your students are frequently anticipated.* Well, I've got one right now with a very fine problem, and someone in the East published two papers and knocked the props right out from under him. . . . But it's good for that student; he's got to think some and go off in some other direction. It's good experience. I look at it from the point of view of value to the student, primarily.

Note the partial inconsistency in these remarks: the informant asserted that anticipation did not prevent publication, but then went on to point out that it necessarily led to a change in the direction of his student's research. On the other hand, this same scientist reported a case which shows that being anticipated in reporting a single result may be of little significance in the course of a large program of research. It concerns an important discovery that gained him considerable fame:

> I would like to say that we published the first paper on this subject which had ever been published. But talk about the influence of editors: we submitted this first paper to *Science* about April, and the editors wrote back to us to say that 'We're sorry, the journal is very crowded and we have room only for the most significant contributions.' . . . In September of that same year a paper on the same subject by a friend of mine in Texas appeared—in *Science*. So we got cheated out of our initial priority. But that doesn't bother

> me any. . . . *Your work on this subject lasted ten years. During that time did you pick up many competitors?* Yes, but we had a head start on them. They would come along and in two or three years they would come out with a paper, and by that time we would have six more. So we really got the jump on everybody. Even the fellow in Texas never caught up with us. He's a nice guy, still one of my good friends.

The severity of competition is reduced when replication is desirable and when results accumulated over a longer period are regarded as more important than spectacular single discoveries.

In mathematics, competition tends to be less prevalent because of the absence of agreement on the relative importance of mathematical problems. Thus, mathematicians may almost welcome anticipation; one said that he was never concerned about the problem:

> I say, bless your heart, if you can get the same result as someone else: it proves the importance of the problem.

This man had never been anticipated, and this partly accounts for his lack of concern. For, although the problem may be less prevalent in mathematics, it is more severe. Replications are not needed, and being anticipated usually forestalls publication. A number of informants told of students who had to begin again on dissertations after having been anticipated when they had almost completed work. The same delays may occur among established mathematicians:

> Once I made the mistake of not publishing abstracts first. I had stored up many results when someone else published a big ninety-page paper which included two-thirds of my results; it wasted two years' work. . . . It came out in a way which surprised me. I knew the man was working in the same general area but I didn't think he could do the same thing. . . . There were also a couple of smaller cases.

Sometimes, however, new proofs for established theorems will be valued by mathematicians, and those who offer them will be recognized:

> The same theorem can be proved in different ways, and the differences may be interesting. Mathematicians are increasingly method-conscious. . . . A one-half-page proof is considered quite an advance over a twenty-page proof. . . . It's not only style. A short proof is likely to be better in other ways; it may be more revealing of other things.

Partly because the same theorems may be proved in many ways, formal scientists frequently rediscover established results. While such rediscoveries may be publishable, usually they are not. They are made more prevalent by the difficulty of finding them in the mathematical literature: the mathematician can often look to see whether his problem has been solved by others only after he has solved it himself. One mathematician reported that being anticipated was "almost the rule" with his work:

> I don't like to study the literature too much. . . . Before getting results one doesn't know where to look; after getting the results you do. If you have a nice conjecture there is a 50 to 60 per cent chance that it's false or useless. And when you do prove it there is a 50 to 70 per cent chance that others have done it.

Such an experience of anticipation is related to competition, and has been discussed here because of this, but it is clearly not the same social form of competition. The "competitors" may have written many years ago, and there is no possibility of a race with them or of devising ways in which to win the race.

As in other fields, competition in the formal sciences may pertain not only to particular discoveries but also to their later development. One mathematician was very proud of a seminal paper he had written on a problem relating mathematics and logic. The development of the ideas he set forth was quickly taken up by others, and he ceased working on them. This was largely because he failed to see the implications of his own work, a fact he regretted.

Some of the variations in the prevalence and severity of competition for recognition within the various scientific disciplines have now been described. Competition is most prevalent and least severe in the physical sciences and least prevalent and most severe in the formal sciences, yet the differences within each area are at least as great as those between areas. Each field includes subjects regarded as of great importance and in which many scientists work, and each field includes subjects of peripheral importance or subjects in which competition is limited by the scarcity of equipment or skills. Nevertheless, competition for priority is a characteristic of all the established sciences. It results from the existence of a system of social control through recognition, and it has behavioral consequences.

Reactions to Competition

Ideally, competition for recognition in science should function as competition for profits in ideal economic markets. First, it should lead to a selection of the most efficient techniques. Scientists may resist the introduction of novel techniques,[14] yet the success of these techniques in achieving important research goals generally leads to their supplanting or supplementing existing techniques.[15] Second, competition should lead to an optimum allocation of research efforts. The research that is valued most highly, viz., the research having the most important results, will tend to attract many scientists, including many of the best ones.[16] This is especially true when a new research area is just opening up, when the problems and techniques are fairly clear and the number of competitors small. At such times, many scientists are said to "get on the bandwagon." (This expression was usually used with disapproval by scientists who did not quickly move into the newer areas.) Some scientists become facile in moving from one newly developing area to another; a number of informants referred to this practice, again with mild disapproval, as "skimming the cream" off the new area. Such practices obviously require special skills, and the scientists who adopt them gain considerable recognition, even if it is accompanied by some envy.

The intense competition in areas of central importance induces scientists working in them to attempt to differentiate their research products and gain something of a monopoly over a small sector of the area. For example, a mathematician working in algebraic topology, an area now considered to be of great importance, said:

> The idea is to find a problem with a gimmick others don't have. Each person has some techniques and is familiar with some aspects of the problem. Many times one mathematician can solve a problem that another cannot.

In the experimental sciences, such product differentiation is often easier. Measurements can be made in different ways and in slightly different ranges, and all will be valuable. Thus, a physics student, when asked if he feared being anticipated by others, replied:

> This has almost happened. I've studied these problems at 230, 260, 290, and 317 million electron volts (m.e.v.), and the Russians have done it at 240, 270, 307, and 333 m.e.v. I've just found out about their work. Actually I think I'm getting more exact answers than they are, but it doesn't matter too much.

Nevertheless, when large numbers of scientists are at work on a relatively small set of problems deemed to be of the greatest importance, product differences of a minor sort are unlikely to lessen concern about being anticipated. The intense competition in areas of central importance forces many scientists to work in areas of peripheral importance. Sometimes this is a reasoned choice. One molecular biologist, for example, was working in the area of protein synthesis, but characterized his own work as:

> a back door approach to it, which may not be a door at all but a blind alley. . . . I think if there weren't any other factors, I would have gone into protein synthesis directly. The reason I didn't is because it is such a popular field, with so many people working in it, and, particularly, rather large laboratories where they have a number of people working on this. I just felt that I didn't want to get into a problem like this. I like to be a little more relaxed about what I do.

This scientist was working in a relatively new department, which had as yet relatively few graduate students; no postdoctoral fellows were working with him. His choice of problem was affected by the research facilities available and, probably, by his evaluation of his own abilities. For the most part the performance of research in peripheral areas is less the result of a decision than the absence of one. Many students, as has been suggested, work on peripheral problems for their doctoral dissertations, simply because students work more slowly and usually cannot meet the competition on problems of central importance. Upon receiving the doctorate, there is a natural tendency to continue research in the same area. Thus, most of the scientists interviewed who were working in areas of peripheral significance were merely continuing in the area of their graduate work. Sometimes they considered moving to areas of greater importance. A young theoretical physicist said, for example:

> My greatest hopes are not so much in my current work but in the most fundamental problems, in particle physics and field theory, but I haven't done very much with them. . . . It would take one or two years to get to the forefront of research. . . . The chances [of being anticipated] are certainly great in particle physics, since so many people are involved and communication is very rapid. *Is this a reason for not going into particle physics?* In a way. It's harder to get satisfaction. The yield per idea is less. . . . One frequently does find that others have had the idea.

Competitive pressures assure that less popular areas of research will not be neglected and thereby facilitate the allocation of work in science.

Competition also encourages taking risks. This is important, for, as will be shown, many elements of the organization of research in universities discourage taking risks. Taking risks is one way of getting around competitive pressures; if successful, considerable recognition is accrued, but the risks involved will discourage others and reduce the competition. A number of informants had "gone out on a limb." Four of them had previously worked in well-established problem areas of central importance and had then selected problems that had recently emerged; their work, if successful, would be of great importance. Such scientists tended to be young men in secure positions, either associate or assistant professors more or less assured of tenure. Failure for them would not be catastrophic, while success would be of great value in their careers. The only scientist in my sample who lacked a secure status, and who gambled and failed, left his university position soon afterward. He was a statistician using a new approach to a relatively old problem:

> Nobody has looked at this as a general problem in estimation theory. . . . I don't have any competitors. This is more or less the academician's ambition, to have an area entirely to one's self. I got into it more or less by accident. *How long do you expect to remain in this area?* I'm not having much luck in trying to generalize this—I can't prove something I'm sure is true. I'll give it another month and then drop it. This will be too bad, since I have much time invested in it.

In this case, risk-taking jeopardized the scientist's tenure in his job—his formal status. In general, jeopardy to formal status is not as important in discouraging risk-taking as jeopardy to informal status—to prestige in small groups of co-workers. Young men, especially when new in a department, will be less likely to be incorporated in groups of co-workers, and this social isolation means they have less informal status to lose and can thereby take risks more readily.[17]

We have seen some of the ways in which competition for recognition in science may lead to a better allocation of research efforts. The most highly valued research, which is that research judged most likely to lead to fundamental consequences for science and technology, receives the most recognition, and scientists are encouraged to perform it.[18] Nevertheless, other research is not neglected; competitive pressures induce many scientists to work in areas considered less important, although of some importance. Calculated risks are encouraged, for, as in economic markets, success in risk-taking results in large rewards. The argument that competition benefits science by leading to an efficient allocation of effort is necessarily tentative and speculative. It rests on a number of assumptions that are difficult to prove. Chief among these are that the rate of development of a scientific field is proportional to the number of scientists working in it and that a law of diminishing returns does not come into play, at least not in the early stages of the development of a field. Neither of these is necessarily true, and the second may often be false. Evidence on these points would seem of central importance to the understanding of the organization of science. In any case, a weaker generalization can be made about the functional nature of competition for recognition in science: it does assure that important areas of research will not be neglected.

Even if the argument that competition ideally leads to a desirable allocation of efforts in science is generally valid, the process does not always have beneficial effects. The allocation of research efforts may be faulty, and deviant means may be used to achieve recognition. First, scientists may engage in "restraint of trade" to restrict competition. This will be discussed in the following section, along with deviant responses. Second, the "supply" of research is typically "inelastic." Unlike actors in economic markets, scientists are under little compulsion to cease producing an undesired product. Furthermore, they may be unable to produce highly valued research, because of the lack of facilities, skills, etc. Third, the scientific reward system may be characterized by "imperfect markets." At one extreme, a few leaders or research administrators may play a large role in influencing the research done by others. At the other extreme, communication barriers may exist that make it difficult for scientists to know what kind of research will bring recognition; when recognition is scarce, it will cease to be an important means of control. These factors will be discussed in Chapter IV. Let us now turn to the first set of factors that limits the role of competition in the control of research—deviant practices and the restraint of trade.

DEVIATION AND ITS CONSEQUENCES

Because science tends to be conceptually unified, and because most important work is eventually replicated,[19] serious forms of fraud, such as the fabrication of data, are rare in science.[20] Fraud is more credible in "field sciences," such as ethnography and geology, than in the experimental and formal sciences, which are being considered here. In any case, the fabrication of data is more likely in peripheral areas of research, where attempts at replication are less likely to occur. Because it is concealed, data about it are difficult to obtain. The following instance, second-hand and therefore suspect, illustrates the possibilities, however. It was recounted to me by a friend of the victim:

> A friend of mine, X, was working for a Ph.D. at a large chemical engineering department. His professor had very few students and wanted more: he was ambitious and had hopes of becoming a university official. X began working on a dissertation based on a dissertation recently completed in the department by a student of an established professor, one who had many students. (The ratio of students to faculty in the department was quite low.) X hoped to extend the other's theory, which was concerned with extractive processes.
>
> After some time of work without success, X found that he could not extend the theory, because it was false. In fact, he concluded that the theory could only have been presented if the other student had falsified his data.
>
> X got his dissertation committee together to tell them that he could not proceed with his research as planned because of this. The situation was socially awkward. Apparently the committee refused to concede that the earlier thesis had been based on falsified data—this would have reflected upon the established professor. They apparently explained X's lack of success as a result of his "incompetence." They may also have been motivated by a desire to put down X's pushy professor.
>
> After some time, X became discouraged and went to another university.

Here, the alleged fraud was perpetrated by a student—a distinctly marginal scientist, and the problem apparently had no central significance to science. Because of this, those involved in the situation could look the other way. This is probably a common response to errors, accidental or intentional, in the empirical sciences. A theoretical physicist said:

> Someone else did some work on a problem I was interested in which I would challenge. . . . I haven't gotten around to it. Work rarely gets challenged; it is often superseded or passed over, but not challenged. Wrong things get done but people ignore them rather than challenge them.

Denying that norms have been violated is one way of maintaining them.

The theft of ideas is a more serious form of deviation since the reaction to it, secrecy, harms science. Plagiarism, the theft of published ideas, is rare because it is so easily discovered.[21] But property rights to unpublished work are not—and cannot be—so firmly established. Since different scientists frequently make the same discovery independently, it is often difficult to substantiate suspicions of plagiarism. (This is also shown by the histories of priority disputes in science where the charge of plagiarism has been raised, such as the Newton-Leibniz dispute on the calculus.) Nevertheless, most scientists are familiar with anecdotal accounts of the theft of ideas, and the conditions for suspicion are often present. Sometimes scientists joke about it. The mathematician-turned-humorist, Tom Lehrer, sings a ditty about a Russian mathematician who publishes the results of others, which he obtains through a very long chain of contacts, and who becomes famous instead of the unsuspecting original discoverers. Usually the reaction to the theft of ideas involves hostility. An organic chemist, in the course of criticizing an English chemist, presented the following episode:

> X has been known to wipe you out if he finds the direction of your research. He will quickly get into the area and publish the most important results. He once scooped Professor Y of this department. Y had talked freely about his research plans, and we are quite sure that someone he talked to visited England and got the news to X. X doesn't usually write up research notes, which are quickly published, but rather substantial articles, which take longer. On this occasion he did, however, lending substance to the idea that he knew Y was working on the same thing.

What is stolen in such cases are ideas, general information about the goal sought and the means to be used. Usually the thief must do considerable research himself. A mathematician may learn about a theorem someone else has proved and the general techniques used, but to steal the idea he must still produce detailed results. A chemist will learn about an experiment that is planned, but he is obliged to do the experiment himself, if he is to publish first. On the other hand, individuals may steal unwittingly.[22] A mathematician was aware of this:

> In some cases, the work hasn't seemed to be entirely independent, and it's possible that someone was unfairly deprived of credit or given credit. . . . I remember one case where a person received a suggestion from someone else, then must have remembered it as his own idea and published it as such.

It is probably easier to steal ideas in the theoretical sciences, but the consequences of thefts may be more serious in the experimental sciences, where large research efforts are involved.

The theft of ideas is probably uncommon and, in any case, has few inimical *direct* effects on science, especially when the thief does a large portion of the research himself. It is even possible that thieves, by publishing ideas which have not been published by their originators, aid in the dissemination of ideas. The serious effects of the theft of ideas are indirect; the possibility of theft tends to make scientists secretive and it leads to the development of various kinds of property rights over unpublished work.

SECRECY

Scientists who are concerned about the possibility of being anticipated as a result of the theft of their ideas tend to be secretive.[23] An organic chemist said he only communicated in detail with persons he was friendly with and could trust:

> I avoided talking to this fellow from [another university] about my work on X compounds. This fellow has a whole group of students, I have only one student working with me. If he wanted to, and if he knew what I was doing, he could get into the area I was working in, clean up the problems, and wipe me out. There is no sense in directing him to the things I plan to do, while it's O.K. to tell him what I've done. After you get to know people you might be able to trust them, and then you can talk freely about your plans.

A biochemist said his behavior was similar: "I engage in correspondence with others. But this is a competitive field, and you don't tell everyone everything you're doing right away."

An experimental physicist said that he discussed his work "very little," even with members of his own department:

> Occasionally I'll ask about a theoretical problem. You can't get very much from such encounters. The person you ask will be interested in other things, and you don't want to tell him too much—all one needs is an idea, and he can go far.

This last individual has already been cited as one who had apparently been anticipated twice in his career on relatively important things. It is possible that his secretiveness and his feeling of being deprived of credit are rationalizations for a relatively unsuccessful career; even so, the rationalization takes the form of fear of further thefts and results in secretive behavior.

Scientists may keep their work secret not only to prevent the illegitimate use of it by others but also to prevent its legitimate use. That is, they may refrain from publishing their work in small pieces as it proceeds, for fear that others will elaborate on it, making it impossible for them to present a complete and coherent piece of work, as has already been indicated. They prefer to keep their work secret until a large effort is completed and then, as some informants said, "drop a bombshell" on their colleagues. Norbert Wiener referred by implication to such behavior among mathematicians:

> I have not sought to work in the profoundest secrecy and to spring my new results on a world which has not even known that I have been working on them. . . . I have never tried to steer other investigators away from my own work so that I could be the beneficiary of the surprise effect of a new paper carefully guarded until I could present it with a maximum impact.[24]

Secrecy as a response to the illegitimate use of one's ideas shades over into secrecy as a response to the legitimate use of them.

Not many informants confessed secretive behavior. This is partly because such secrecy conflicts more or less with the norms of free communication in science; secrecy itself must usually be kept secret. It is therefore difficult to form estimates of its prevalence. Scientists within the same specialty often made widely divergent estimates of the prevalence of secrecy. For example, a molecular biologist of wide experience claimed there was not much secrecy in his field:

> There is an element of this, but not really so much. . . . I think there is an astonishing amount of collaboration between people who might well be thought of as competitors. I think we have grown up a great deal in this regard.

An associate professor in this molecular biologist's department thought differently:

> There is a certain amount of secrecy among people, which varies from individual to individual, particularly if you know that somebody else might be interested in your work because they are doing something similar. There is a tendency then not to talk about it too much. . . . Some people are very free in what they talk about. At least they give you that impression. And others are not so free. *Is it provoking to find someone whom you think is working on the same problem you are and who won't talk?* From my own experience, what is more annoying are certain individuals who will ask all sorts of pointed questions about your own work and will absolutely refuse to say anything about theirs. I think if it is an exchange, this is fine. But with this other aspect, where you feel that one person is taking advantage of you, it is different.

In another case, an experimental physicist who was chairman of his department reacted in the following way to the suggestion that some scientists are secretive in response to competitive pressures:

> I don't think so. I don't notice it much around here. None of my colleagues seems to be worried about that, and the general atmosphere is one of eagerness to discuss what they are doing with visitors. I've never noticed any of them withholding any information at all.

His associate, who occupied the neighboring office, was quoted earlier as saying, ". . . . you don't want to tell [your departmental colleagues] too much—all one needs is an idea, and he can go far." It is possible that the individuals who strongly denied the existence of secretive behavior were themselves not at all deceived but were attempting to deceive the interviewer, in order to present their discipline in a favorable light. Whether or not this is true, their remarks show that secrecy among scientists is a form of deviant behavior and indicate the difficulty of measuring its prevalence.

Secrecy may result from other elements in the situation, in addition to the fear that ideas will be stolen. Scientists may be cautious, afraid to present their ideas publicly until certain of their validity. Charles Darwin, who kept his evolutionary ideas secret from all but a few close colleagues for twenty years, is a notable example. Some of my informants asserted that secretive behavior they had observed had this justification. A mathematician said he thought mathematicians generally did not conceal results for fear of being anticipated:

> Not the people I talk to. They are very open about discussing their work while it is in a formative stage. Reticence stems not from fear of being anticipated but from fear of being not right. It is embarrassing to be wrong. One would rather be anticipated than to tell people important results which are wrong. Mathematics and the experimental sciences differ; in the latter you expect approximations to the truth. An error of 10 per cent may be O.K., but this is not so in mathematics, because of its very nature.

While it is easy to see how such an argument applies to formal communications, it is not so easy to see how it applies to informal ones, in which it is customary to warn the speaker that the results are not finally validated.

Although it is difficult to generalize about the relative prevalence of secretive behavior, many informants agreed about the following propositions. First, secrecy is greater in Europe than in the United States, and it is greater in the eastern U.S. than in the West. For example, a biochemist said:

> You get the feeling [that people are secretive] in some parts of the country. I haven't got this feeling out West, but you do in the East. Everyone who comes out here gets this feeling of freedom and informality—it's one of the first things that impresses them. In the East you often don't say anything about an experiment until the paper is sent in.

A German-trained mathematician thought there was less secrecy among American mathematicians than among those in Germany:[25]

> At Göttingen everyone hid his ideas until they were mature. . . . I was very encouraged to see that people are quite free in discussing in this

> country. This is partly because competition in my field is very low. In Göttingen, when I got a job as an instructor, the lowest rank, I had about thirty competitors.[26]

This informant suggested, as an explanation for varying degrees of secrecy, that it was linked with competition for position.

The second proposition suggested by my informants about the relative prevalence of secrecy is that it is fairly common in organic chemistry and biochemistry, very rare in the formal sciences, and somewhere between these extremes in the fields of physics and molecular biology. In part, the prevalence of secrecy may depend on the severity of the consequences of being anticipated. It is also affected by the research facilities required. Since the facilities in high-energy physics are expensive and shared by many physicists, decisions about which experiments are to be performed tend to be collective, at least in part, and are almost necessarily public. Secrecy is almost impossible.[27] However, neither the severity of competition nor differences in research techniques accounts for all the differences in secretive behavior between disciplines; to some extent these differences are customary and self-sustaining. Secrecy breeds secrecy, and openness breeds openness.

Secretive behavior has both personal and social costs. It tends to isolate the individual, and this is usually interpreted as a deprivation. As one experimental physicist (solid-state) said:

> Lots of people keep their work secret. I find that kind of situation unbearable. If you do this, you can't expect others to communicate with you. . . . I prefer exchanging information to producing the whole thing *in toto* before others think of it. I know many people who are anxious about producing the whole thing. They tend to get ulcers thinking about it.

A theoretical physicist thought it led to lower productivity:

> [Secrecy] is hard—there is too much communication. One would have to be hypocritical, consciously secretive. It would spoil the fun of being a scientist. Anyway, it is hard: if you refused to tell a visitor your results he would find out by talking to the man in the next office. I can't imagine such holding back happening except in really isolated cases, which are almost nonexistent. It has been proved that you benefit by discussion.

This personal cost, lower productivity (and, thereby, lower recognition), is also a social cost.

Secrecy has other social costs. It leads to a duplication of effort that might be avoided. As has been noted, it may lead to a reduction in formal communication: the scientist may seek to present only results and withhold his methods. This restricts the dissemination of new and possibly important techniques, and it also makes for greater difficulty in determining the validity of results. Finally, secrecy results in less solidarity for the scientific community as a whole and for particular work units within it. Reduced solidarity increases the probability of deviation of all sorts—restricted production, fraud, loss of objectivity, and so forth. Various practices have emerged to reduce the individual and social costs of secrecy. These practices may themselves have unanticipated and detrimental consequences, however.

OVERCOMING SECRECY: DIVISION OF THE PROBLEM

To the extent that scientists can establish property rights over work in progress, they need not fear anticipation, and the need for secretive behavior is reduced. Such property rights may be more or less explicit and formal. In many scientific specialties, scientists are personally acquainted with all workers in the area. When it becomes evident to two of them that their research will probably produce the same results, they may informally agree on a division of labor, with one scientist producing part of the results and the other the remainder. When the results are logically interdependent, this necessarily results in collaboration, the joint authorship of results, but frequently the results are not logically interdependent and can be published separately. For example, an organic chemist kept in close contact with others in his field, especially with those in the department where he did his doctoral research:

> We are communicating with all those working [on X compounds]. . . . I know the people at Y University. Our communications essentially lead to a division of the problem.

Such agreements to divide the problem can be expected to occur only between individuals who are known to each other and who trust each other. Before a man can be considered for such an agreement, he must have shown possible competitors that he can compete and can be trusted. The publication of good research is the best way of demonstrating the former. The latter probably is demonstrated by being associated with others in the same institution, as with teachers and their former students, or by a kind of introduction through third parties well-known to both competitors. Sometimes acts of deference demonstrate a man's trustworthiness. An experimental physicist who was an international leader in his specialty described the following case:

> Sometimes we send out technical reports a year ahead of the appearance of things in the journals, and we talk to people on the telephone even before that. I know what 80 per cent of the people in the area are doing. In the early days, before I was well known, I visited Europe and saw X in England. I showed him some of my unpublished work on ammonia, and he showed me some of his independent results on the same thing. I waited until his

appeared in print before submitting mine. He referred to my unpublished work. I was ready to publish first, but I deferred to his authority. His good will was more important to me than priority.

The division of the problem may be more explicit and formal when expensive and long-term research efforts are involved. For instance, physical laboratories having high-energy particle accelerators may agree more or less openly on a division of the problem. (Of course, they may agree to compete.) Several astronomical observatories may divide up the heavens for purposes of mapping stars. Perhaps the most extensive form of such organization took place in the International Geophysical Year of 1959–1960, when thousands of scientists made different but related observations of geological and meteorological phenomena. Such extensive co-operation takes place mostly for complex data-collection tasks, rather than for tasks of a more theoretical nature. Even then, scientists must be willing to accept some directives from a central planning body. In general, contractual relations, viz., agreements formed and agreed to by independent agents, exist only between pairs of individuals, or, infrequently, among three. When more parties are brought in, difficulties of communication arise and it becomes more difficult to find a way of dividing the tasks and rewards agreeable to all parties. Therefore, a division of the problem among more than two or three research groups requires either some formal planning agency or some form of informal leadership. Both possibilities infringe on the norms of autonomy for the individual scientist.

It can be seen that a division of the problem is likely to include only a few scientists in a specialty. Usually this will not significantly reduce competition, and pressures toward secretive behavior will continue to exist.

OVERCOMING SECRECY: EARLY PUBLICATION OF PARTIAL RESULTS

In the year 1610, Galileo sent the following announcement of an astronomical discovery to the Tuscan ambassador in Prague:

SMAISMRMILMEPOETALEUMIBUNENUGTTAURIAS[28]

Galileo was not prepared to publish his complete set of observations or his analyses of them, but he wished to assure himself, through publication in code, that his priority in the discovery could be proved.

In the United States today, scientists have found different ways in which to claim priority without actually publishing their full results. This is usually done by taking advantage of the opportunities furnished by various periodicals to publish abstracts of forthcoming papers, or letters about recent results.[29] The manifest function of such forms of publication is usually to assist those attending scientific meetings in selecting the sessions they will find most interesting. The Federation of American Societies of Experimental Biology, for example, holds an annual meeting that is the largest of its kind in the country. Prior to the meeting, abstracts of the papers to be presented are published in the *Federation Proceedings.* (This is a large volume: 2,400 papers were given at the 1961 meeting.) An informant said:

> . . . in most instances these are only two hundred and fifty-word abstracts. In most instances these are preliminary statements to be followed by a more lengthy article. . . . They permit people attending the meeting to select the program which they wish to attend. I think publishing abstracts probably serves the purpose of a certain amount of claiming of a procedure with some results. It undoubtedly serves the purpose, but this wasn't the original reason at all. It was just to let people know what sort of a program was coming.

The latent function of publishing abstracts is to permit individuals to "stake a claim," establish property rights on research in progress.

Almost all the disciplines in the formal and experimental sciences possess something similar to the *Federation Proceedings.*[30] The American Mathematical Society holds frequent meetings in many parts of the country, and abstracts of papers delivered at the meetings are published in the *Notices* of the Society. The Society has also adopted the procedure of submitting papers to the meetings "by abstract only"—the abstract will be published even if the author does not attend the meeting in question. The American Physical Society also publishes abstracts of papers delivered at its meetings and, in addition, abstracts of papers to be published at some future time in *The Physical Review.* The Physical Society also has a means for more rapid publication of results in *Physical Review Letters.* This is issued every other week and letters appear in print very soon after they are received. It was originally founded as a means of rapid communication of important results that would be of general interest to physicists, rather than as a journal in which specialists communicate to other specialists. However, the possibility of rapid communication has tempted many physicists to use the *Letters* as a means for staking a claim in an area and anticipating their colleagues. It has published a number of editorials critical of this practice:

> . . . one of our ticklish problems concerns the large number of contributions that pour into our office when a "hot" subject breaks and many groups initiate related work. Examples are parity nonconservation, maser developments, and most recently the Mössbauer effect. Because of the high current interest, the rapid development, and the intense competition, we have found it necessary to relax our standards and accept some papers that present new ideas without full analysis, relatively crude experiments that indicate how one can obtain valuable results by more careful and complete work, etc.—in short, papers which under less "hot" conditions would be returned to

authors with the recommendation that further work be done before publication. . . .[31]

Our contributors have very strong biases—concerning what is important, urgent, and promising, and concerning the value of the field of interest as compared with other branches of physics. The referees whom we consult in cases of doubt have similar biases. . . . Occasionally it becomes necessary to ask advice from more than one expert about a particular contribution. . . . With the rapid exploitation of ideas, priority questions become serious problems. . . . We do not take kindly to attempts to pressure us into accepting Letters by misrepresentation, gamesmanship, and jungle tactics, which we have experienced to some (fortunately small) extent.[32]

In another editorial, the Editor records some of his "pet peeves":

. . . an author who uses the "Letters" merely to announce a later paper and whose letter is incomprehensible by itself; an author who publishes a "Letter" which is merely an amplification of a previously published meeting abstract; . . . an author who carries a chip on his shoulder and casts aspersions on the motives and integrity of a referee who gives an adverse report on his paper; an author who tries to sneak a Letter in to "scoop" a competitor who has already submitted a full Article. . . .[33]

Scientists frequently forget the manifest function of forms for early and prior publication of results. What was a latent function, the opportunity to claim some property rights on research in progress, becomes the manifest function.

Although many scientists refer to the publication of abstracts, notices, and letters as "staking out a claim" on a problem, the extent to which such claims are honored and the reasons for honoring them are unclear. Some scientists apparently feel that partial publication is equivalent to publication in an article; even if the result is published without extensive proof, it is felt to establish clear priority. In this connection, a theoretical physicist said the following about scientific meetings: ". . . the purpose of meetings is to get your abstract out. You give your talk, then you've got your name on it."[34] Nevertheless, the methods of proof are an indispensable part of scientific results. If one individual gave an abstract and another followed quickly with a complete proof, the latter might claim some priority. Thus, a logician said: "[An abstract of a result] is sort of a claim on it. It's not clear just how much of a claim, since he can't give all the details in the abstract."

In the experimental sciences, where competitors will have invested much time and effort in an experiment, there is a far stronger tendency to attempt to publish the first complete paper, even if another group publishes an abstract first. This tendency is inhibited by the likelihood that the group publishing the first abstract will also be able to publish the first full paper; if so, the investment of further time and effort would be without value. Sometimes competitors make personal contact with one another at this stage. A biochemist gave the following example:

> Very often, in biochemistry, people will be thinking along the same lines and start on the same idea at the same time. If somebody beats them to the punch in publication, very often you see them after awhile at a meeting, and then you sit down and see what each lab is doing. . . . For example, we published a paper on X a few years ago, and it turned out that a group at Y university had done the same thing, got the same results, but didn't publish; we were about six months ahead of them. That often happens. We talked about it at a meeting, and they went along some other lines, although still interested in X. In most of these things there's room for everybody. If someone wants to take credit for everything, usually he loses out in the end.

In general, however, more than mere prudence—the feeling that the individual who first publishes in an abstract will eventually be given credit for priority—is involved. Usually there appears to be a specifically normative component, a feeling that it is not proper to publish a paper on a result already suggested in an abstract. A biochemist stated this in the following way:

> The usual system is to write a short note and send it to one of the journals which publishes notes. It gets into print in a month or so, and then you can take your time about a big long paper. . . . *A short note in a way establishes your claim to the area?* Yes. You've stated what you found, and then someone else can't come along and say that he found it first. . . . *Someone else might come along and write a long paper on the subject—or isn't this done?* Not in the best of circles. I wouldn't do that without confiding in the original author, telling him what I'm doing, asking him if this would conflict in any way with his interests, and even offering to send him the manuscript ahead of time. Most scientists, I am happy to say, are rather honorable people.

Most, but not all. The mathematician Norbert Wiener has written a detailed account of his misbehavior on a similar matter, in which property rights were not established by the publication of abstracts but rather more informally. Early in his career Wiener was informed of some new problems in the area of potential distribution by a Harvard professor whose students were writing dissertations in the area. Not knowing, he writes, "how carefully many professors conserve problems for their own graduate students and how sharply they regard proprietary rights in new problems," Wiener began to work on the problems and had soon gotten further than the doctoral candidates. Although he was warned that it would constitute a serious breach of manners, Wiener published his results. He reports that he was severely criticized for doing so.[35]

In general, however, considerations of morality and prudence lead scientists to respect the claims to results established by publication of abstracts. In this way, when a scientist has reached the point at which he can publish some partial results, the competitive pressures confronting him are reduced. The property rights offered by the opportunity to publish partial results quickly obviate much of the need for secretive behavior. But, while such procedures may facilitate open communication in science, they have other consequences that are dysfunctional. The recognition of such claims tends to discourage replication of the work, which is important if the results have wide implications. Furthermore, since methods and proofs are not described in detail in abstracts, readers cannot readily judge how much confidence to put in the results.[36]

The problem of being unsure of results published only in abstract form is found in all fields, but it is especially serious for such fields as molecular biology, in which the degree of precision and control in experiment is low and scientists often disagree about how to interpret an experiment. While scientists in all disciplines had doubts about the value of publishing abstracts, the molecular biologists interviewed expressed the strongest opinions. For example, a biochemist said:

> . . . to get a short note like this published, you generally have to present some experimental data, but you don't really show the whole controlled experiment and all that really establishes it. Then, two years later, you still haven't done so.[37] This leaves everybody in the field in quite a quandary. Here is a result, a conclusion has been published, without adequate background. What can you do? You can't go ahead and assume for sure that it is true, yet you don't want to go back and repeat the work, for someone has already done it and is presumably writing it up. You can write to him and ask, "How good is the evidence for this and this?", and he may or may not answer you or tell you. It seems to me beyond question that this matter of publishing preliminary notes is more bad than good for science. It may be more good than bad for the individual scientist.

Another biochemist felt that abuses of this sort could be controlled by editors of scientific journals:

> Sometimes people . . . will publish something in a hurry that they will later have to retract or that they will never follow up because they cannot confirm the original observations. Now that type of thing—it's all right, but I don't think it's a legal reference. Before long I don't think most editors will permit references to abstracts in full-length manuscripts, if they are quoted as documentation for a certain point.[38]

That is, the editor would not permit authors of manuscripts to recognize the priority, the property rights, of those who have published in abstract form only.

From the perspective of the deviant—the scientist who publishes an abstract without soon following it up with a complete paper—the situation appears less serious. A mathematician who had done this gave the impression of challenging his colleagues to provide a proof and solve a problem:

> Occasionally I have held up publication of something for a long time to see if someone else could do it. In one case I simply stated a result in an abstract fifteen years ago, and although in these fifteen years others have used the result, no one else has been able to fabricate a proof for it. *Maybe they considered it to be your property and were waiting for you to publish?* Mathematicians never feel that way.[39] If a proof hasn't been given, it's a fair domain for everybody. . . . It is true that, since I stated the theorem first, I get some credit for having formulated it, but until a proof has been published you don't have any more position with regard to that proof than anyone else who comes along and proves it. . . .
>
> I didn't do this intentionally. I felt that the proof was not in its best form and hoped to get back to it and get it polished up, but I got interested in other things and never got back to it. Finally, a student of mine, who was working in a related area and used the theorem, published the proof as part of one of his papers. Maybe it wasn't too fair a thing to do. I had some communications from some people abroad who were interested in the proof and I had nothing to send them; I gave some general ideas in letters but I don't know whether or not they were able to follow it out. I don't think this is a good thing to do. It is important to get proofs out. If I had thought that the results were important enough so that a great deal of theory was built up on it, I would never have held up like this. It was an interesting result, but not one to give rise to a whole body of theory.

Even this informant became somewhat defensive when it was suggested that the publication of results without proof might have disadvantages, and he asserted his desire to conform when his results were of considerable importance.

Some ill effects for science may arise simply because scientists are uncertain about rules of behavior in this area. Scientists who believe that publication of abstracts gives them proprietary rights on results will tend to treat abstracts as serious communications, will hasten to publish complete papers with supporting evidence, and will respect the proprietary rights of others. Scientists who believe that abstracts do not give proprietary rights will be less ready to treat them as serious communications, less ready to publish complete papers, and less ready to respect the proprietary rights of others. When both attitudes are prevalent, some scientists will defer to the rights of others and assume them to be responsible; the rest may feel they have no special rights and fail to assume the obligation of publishing complete proofs. If either norm were firmly established, the existence of published results of doubtful validity would not be such a problem.

The scientists in any discipline have the power to prevent the establishment of claims to results by the publication of abstracts; they need only refuse to recognize such claims. (The publication of abstracts could still be encouraged for other reasons, of course.) If claims to results are not recognized, the problem of how to deal with what may be invalid results in the literature is partially solved. However, this solution means the abandonment of one device to control competitive pressures. Without such devices scientists will be more likely to engage in secretive behavior, which is itself disadvantageous to science. The dilemma between recognition of rights of priority and the maintenance of open channels of communication can be softened in various ways, but it cannot be completely resolved.

COMMUNITY PROPERTY: FOLK THEOREMS

Recognition of the priority rights of individuals may lead to a situation whereby important results lacking adequate proof are accepted and used by many scientists. The same situation may also arise when scientific results are considered to be community property, property no individual has the right to assume for himself. In mathematics such properties are often designated as "folk theorems." The following complaint about the presence of such theorems appeared in the *American Mathematical Monthly* in 1957:

> Each mathematical discipline needs to draw on the others; yet it is impossible for even the best of us to keep abreast of the research in more than a small part of the field. We are separating off into small groups of specialists with little intercommunication. . . . A published paper is not properly an open letter to a half-dozen fellow-specialists who understand the motivation, possess the antecedent information, and can readily fill in omitted definitions and proofs. An addition to mathematical knowledge has significance as part of the body of all mathematical knowledge, and its place in that body should be clearly indicated. . . .
>
> There are mathematicians (fortunately few) who consider that a speaker has somehow "lost face" if he has spoken so as to be intelligible to any but the select few. One hears of "folk theorems" established (presumably) by some expert, communicated verbally or more likely mentioned in an offhand way during some conversation with another expert or two, and thereafter unpublished forevermore because no one would want to publish a "known theorem."[40]

According to this writer, the use of folk theorems in publications should be restricted, and he encouraged mathematicians to publish proofs for such "known" theorems.

An algebraist said that publication of this sort would be an unjustifiable appropriation of community property:

> One part of keeping up with one's field is knowing the folk theorems. Nothing disturbs me more than seeing someone rush into print with a lot of folk theorems and sort of claim them as their own. That's very bad. When folk theorems get put into print, which is perfectly all right, those

> who do so should not claim personal credit for them. This is particularly true in my subject [lattice theory] . . . a great many people jumped in early and skimmed the cream off the subject, and it was difficult at first to tell the difference between the trivially easy work and other things. The result was that many of the researchers in the field who were interested in getting at the deeper aspects would develop a great many more or less superficial things which were important as technical aspects of the subject but were not of great interest in themselves. And this becomes, if you wish, a body of folk knowledge about the subject, and only a small fraction of this got into print. It finally got to the place where we got homilies like "It is well known that," and "These results can be stated without proof." There is such a body of knowledge. It is gradually coming into print as books and texts are written. That's the way it should come in, rather than through papers by people who have not found the proof in the literature.[41]

In other fields, folk knowledge exists in the form of both experimental techniques that are widely used but not described in print and commonly accepted matters of fact. The latter are most likely to come to scientists' attention when their validity is challenged.

Scientific knowledge is community property. Discoverers have limited rights, but among them are rights to be recognized for their discoveries. In marginal cases, when a discovery is announced but evidence is not put forward in detail, recognition of property rights inhibits communication and increases the probability that invalid propositions will be accepted. Such marginal cases arise most often because individual scientists attempt to reduce competitive pressures by claiming priority before publishing their complete work, but cases may also arise when commonly known results are unclaimed by any particular individual. Common knowledge tends to be regarded as community property, even if evidence for it is not presented publicly. Individuals who attempt to claim rights of discovery for it will be discouraged, for in doing so they steal from all.

Summary and Conclusion

If science were an individualistic enterprise, with each scientist oriented primarily to increasing his own knowledge of the workings of nature and only secondarily to sharing this knowledge with others, the possibility and the experience of being anticipated would have little significance. Scientists would be gratified by having others solve their problems, a gratification mitigated only by the denial of the pleasure of solving the problem first for one's self—a trivial deprivation comparable to being told the conclusion of a detective story while in the midst of reading it.

In fact, however, the possibility of anticipation is a serious concern to many scientists. Scientists prize the recognition that comes from being the first to present a discovery, and they compete strenuously to obtain it.

The prize is of such importance that they may sacrifice other scientific values to obtain it. Fraud and plagiarism may be rare, yet scientists may violate the norms of free communication in science in order to protect their competitive standing, and they may default on their obligation to present only thoroughly verified results in the interest of publishing quickly. Various procedures have arisen in the scientific community to control such deviant tendencies, but they have inherent limits. If property rights are granted not only to finally published results but to problems currently engaging scientists as well, tendencies to be secretive will be sharply reduced and the norms of free communication maintained. But the establishment of such property rights and the concomitant restriction of competition retards scientific progress and induces scientists to present partial results of dubious validity. Attempts to curtail such practices, however, increase the likelihood of secretive behavior.

If the incentives in science were the same as those in the world of business, where rewards of money and power are given for specific performances, competition would exist but it would be of a different order than that now found in science. Applied scientists in industrial employment compete, but priority in publishing their discoveries is relatively insignificant. In industrial employment, being anticipated is of little importance, provided the information is put to use in gaining competitive advantage for the concern, and being first is of little value unless it leads to such advantage. Industrial scientists' publications are typically withheld for periods of as long as a year or more, until discoveries can be patented. (Patent laws usually deny patents on work already published.)

The nature and prevalence of competition for priority among scientists is strong evidence, then, for the validity of the information-recognition exchange model of scientific organization. In this chapter, it has been suggested that this type of organization may be functional because it induces scientists to select appropriate methods and to allocate their activities efficiently. In succeeding chapters, this suggestion will be qualified by distinctive types of disorganization associated with this type of organization. First, however, we shall consider how competition is complemented by the process of collaboration, whereby scientists work interdependently in pursuit of a goal and agree to share the recognition for its achievement.

NOTES

1. It is fitting that many of the points made in the ensuing pages appeared in an article I discovered after the first draft of this chapter had been prepared. See F. Reif, "The Competitive World of the Pure Scientist," *Science*, 134 (1961),

1957–1962. Reif does not consider differences among scientists with respect to competition, and his conclusions are presented without supporting evidence.

2. Cf. the experiences reported by Norbert Wiener in *Ex-Prodigy* (New York: Simon and Schuster, 1953), p. 281; and *I Am a Mathematician* (Garden City, N.Y.: Doubleday, 1956), pp. 92 f.
3. Because of their spectacular nature, simultaneous discoveries have often been noted by social scientists, who have used them as evidence against naïve individualistic points of view. Robert K. Merton pointed out that Macaulay had written about them as early as 1828, and their importance has been noted by others, from Auguste Comte to Alfred Kroeber, down to the present day. "Singletons and multiples in scientific discovery," *Proceedings of the American Philosophical Society*, 105 (1961), 470–486. Despite this, some writers still deny their prevalence. See, e.g., Tertius Chandler, "Duplicate inventions?", *American Anthropologist*, 62 (1960), 495–498. Chandler suggests that in most alleged simultaneous discoveries the real discoverer has influenced or has been copied by others. His evidence is unconvincing.
4. In fact, Merton suggests that "all scientific discoveries are in principle multiples," that only in exceptional cases are discoveries made by a single individual, and he adduces many reasons for this. Merton, *op. cit.*, p. 477.
5. The information was not obtained in thirteen cases, either because the scientist was not productive or because the questions were not asked. Being anticipated probably happened more frequently among the scientists in my sample than for the average scientist because of the bias toward eminence in the sample. Nonproductive scientists, or scientists who select trivial problems, will have less opportunity to be anticipated.
6. One confounding factor may be the relatively high productivity of the younger men in this sample. If the sample is divided into groups of high and of moderate or low productivity, the correlation between productivity and experience of anticipation is $Q = .18$ ($N = 63$).
7. Similarly, concern about anticipation is greater for those with low productivity, since the marginal value of each publication is relatively greater for them. The correlation between productivity and concern about anticipation is $Q = -.35$ ($N = 58$).
8. Sometimes a scientist's concern about being anticipated increases with the passage of time. This can happen when a monopoly of techniques or equipment is broken.
9. *The Higher Learning in America* (New York: Sagamore Press, 1918), pp. 68, 71.
10. Cf. Thomas S. Kuhn, "The Function of Measurement in Modern Physical Science," *Isis*, 52 (1961), 190.
11. Competition will be least prevalent and severe in the social sciences. In sociology, for example, theory is not well enough developed to enable specialists to select problems that make it possible to test theories at critical points. The low precision of sociological methods does not engender in specialists the confidence that theories have been supported or weakened by particular results, and, on all but the obvious points, replication is always desirable. Furthermore, the field is new and the number of workers small.
12. Cited in Michael Polanyi, *The Logic of Liberty* (London: Routledge and Kegan Paul), 1951, p. 50.
13. As with economic oligopolists, scientific oligopolists tend to enter into collaborative (or "collusive") relations. That is, the major laboratories with high-energy particle accelerators tend consciously to divide the available problems among themselves.

14. Cf. Bernard Barber, "Resistance by Scientists to Scientific Discovery," *Science*, 134 (1961), 596–602. (Reprinted in Bernard Barber and Walter Hirsch, eds., *The Sociology of Science* [New York: Free Press of Glencoe, 1962], pp. 539–556.)
15. It is beyond the scope of this study to provide detailed evidence for this proposition. A discussion of the almost inevitable success of quantitative methods, when they can be used, is presented in Kuhn, *loc. cit.*
16. The possibility of self-fulfilling prophecy, analogous to behavior in speculative markets, exists here. An area defined as being highly important attracts the interests of highly skilled scientists; then, because they are highly skilled, they produce important results that confirm the original evaluation of the problem area. It is conceivable that such scientists need not even be the most skilled, but merely those with greatest prestige. The situation would then be analogous to fashion leadership; the leaders of international café society do not necessarily *select* the top fashions, but may *produce* the top fashions by their choices. It is probably impossible to put speculations such as these to rigorous test. They will be discussed further in Chapter IV.
17. For a summary of research on this topic and a theoretical elaboration, see George C. Homans, *Social Behavior: Its Elementary Forms* (New York: Harcourt Brace and World, 1961), ch. 16.
18. For a similar but less sociological analysis, see Gerald Holton, "Scientific Research and Scholarship: Notes toward the Design of Proper Scales," *Daedalus*, 91 (1962), 362–399.
19. Replication seems to occur less frequently in the behavioral sciences. See Theodore D. Sterling, "Publication Decisions and Their Possible Effects on Inferences Drawn from Tests of Significance—or Vice Versa," *Journal of the American Statistical Association*, 54 (1959), 30–34. Sterling examined a year's output of four psychological journals, which included 362 reports of experimental studies, and he found that not one was a replication of previously published work.
20. See Robert K. Merton, "Priorities in Scientific Discovery," *American Sociological Review*, 22 (1957), 649–651. (Reprinted in Barber and Hirsch, *op. cit.*, pp. 447–485.) A fictional case of fraud is presented in C. P. Snow, *The Affair* (New York: Scribner's, 1960).
21. There are exceptions. M. de Duffahel, a French mathematician, republished under his own name some of the classical papers of the great masters. He was discovered by a reviewer in 1936, when he submitted a paper published twenty-four years before by C. E. Picard. Paul R. Halmos, "Nicolas Bourbaki," *Scientific American*, 196 (May 1957), 88–99.
22. This happens so frequently that Robert K. Merton has coined the term "cryptomnesia" to refer to it. (The alternative, "unconscious plagiary" has inaccurate connotations.) See his discussion in "Resistance to the Systematic Study of Multiple Discoveries in Science," *European Journal of Sociology*, 4 (1963), 272–282.
23. This is an example of a general proposition that can be made about competitive situations. Josephine Klein has summarized the results of much small-group research by saying, "In his analysis of co-operative and competitive groups in action Deutsch shows some interesting differences between them. For instance, competing members are careful to withhold information from one another, with the result that members of the co-operative group learn from one another to a much greater extent than do competing members." *The Study of Groups* (London: Routledge and Kegan Paul, 1956), p. 35.
24. *I Am a Mathematician*, *op. cit.*, pp. 87 f.

25. Cf. N. F. Mott, "Working for a Society Where Science Can Thrive," *Bulletin of the Atomic Scientists*, 7 (1951), 375: "It is perhaps the special excellence of science in the Anglo-Saxon countries that this exchange of science is so free. In the United States or in England, if we have some new results, if we have an idea, we go and tell the man in the next room. . . . We aren't jealous of ideas; we aren't afraid of other people stealing them. I have heard visitors from the continent of Europe say that this is what strikes them most about our science."
26. The same informant went on to say, "Some mathematicians publish their results in such a form that others can see the results but not the methods used to achieve them." This kind of obscurantism prevents others from developing the ideas of the discoverer and gives him the opportunity to present the whole corpus himself. Needless to say, it results in poor mathematics: in this field, as in others, the method is often as important as the results. Newton's *Principia*, the results of which were arrived at with the use of fluxions, presented the results in formal and traditional geometric language, without the fluxions. I do not know whether this represents Newton's conformity to traditional techniques or whether it was an example of the same kind of secrecy.
27. Scientists in general face a similar problem when requesting funds from government agencies and foundations; they sometimes fear that members of advisory panels may steal their ideas. Harold Orlans reports: "Occasional incidents have lent some substance to this concern," and some scientists attempt to establish a prior claim through publication and only then submit a grant request. *The Effects of Federal Programs on Higher Education: A Study of 36 Universities and Colleges* (Washington, D.C.: Brookings Institution, 1962), pp. 193, 257.
28. This is a Latin anagram of, "I have observed the highest planet [Saturn] in triplet form." Arthur Koestler, *The Watershed* (Garden City, N.Y.: Doubleday Anchor, 1960), pp. 199 f. Galileo revealed the meaning of the anagram three months later, at the behest of the Emperor Rudolph. He would have been wiser not to, since he was wrong in his interpretation.
29. Derek J. de Solla Price suggests that articles published in scientific journals were themselves developed to perform this function. *Little Science, Big Science* (New York: Columbia University Press, 1963), p. 68.
30. Still other techniques are used in such fields as history and anthropology. "Some historians use the correspondence columns of the *Times Literary Supplement* to put fences around their research questions with the innocent statement 'I am working on a biography of Lord William Bentinck and would appreciate information about any uncatalogued letters received or sent by him.' This is more than a request for information; it is a warning as well. Some anthropologists are also concerned with people working in their fields. But in recent years replication studies have shown more and more the possibility of multiple explanations." Bernard S. Cohn, "An Anthropologist among the Historians: A Field Study," *South Atlantic Quarterly*, 61 (Winter 1962), 20.
31. S. Pasternak, "Editorial," *Physical Review Letters*, 4 (1960), 395 f.
32. S. A. Goudsmit, "Editorial," *ibid.*, 4 (1960), 109 f.
33. S. A. Goudsmit, "Editorial," *ibid.*, 6 (1961), 587 f.
34. The interviewer's question was about abstracts, whereas the informant's answer was about papers actually delivered. The impression given, however, was that publication of the abstract gave priority rights.
35. *I Am a Mathematician, op. cit.*, pp. 79–85. Compare the experience of Thomas Huxley. In one of his letters he described how, in order to publish a paper, he

had to be sure it got into the right editorial hands, rather than those of someone who had property rights in the area. He wrote, "You have no notion of the intrigues that go on in this blessed world of science. . . . Merit alone is very little good; it must be backed by tact and knowledge of the world to do very much." Quoted in J. R. Kantor, *The Logic of Modern Science* (Bloomington, Ind.: Principia Press, 1953), pp. 57 f.

36. Sinclair Lewis described the way competitive pressures lead to hasty publication in *Arrowsmith* (New York: Harcourt Brace, 1934, p. 324):

> "Throw your material together as rapidly as possible and send a note in to the Society for Experimental Biology and Medicine, to be published in the next proceedings."
>
> "But I'm not ready to publish! I want to have every loophole plugged up before I announce anything whatever!"
>
> "Nonsense! That attitude is old-fashioned. This is no longer an age of parochialism but of competition, in art and science just as much as in commerce—co-operation with your own group, but with those outside it, competition to the death! Plug up the holes thoroughly, later, but we can't have somebody else stealing a march on us. Remember you have your name to make. The way to make it is by working with me—toward the greatest good for the greatest number."

But Arrowsmith is anticipated just the same.

37. In fact, the whole experiment may not be published because it has not confirmed the results reported in abstract form. As one informant said, a disadvantage of the procedure of publishing short notes was that "you might have to reverse your field later on when things don't develop right."

38. The editors of *Physical Review Letters* have announced that they will discourage references to unpublished manuscripts such as pre-prints, on the grounds that the author's reasons for not publishing a paper may be good reasons for not quoting it. However, nothing specifically is noted about references to abstracts. George L. Trigg, "Unpublished Referençes," *Physical Review Letters,* 11 (1963), 1 f.

39. Other mathematicians disagreed. Many of them spoke of the publication of an abstract as having "staked a claim" to a result.

40. E. J. McShane, "Maintaining Communication," *American Mathematical Monthly* 64 (1957), 315 f.

41. This is one reason why the publication of texts tends to be a despised form of scientific communication: the textbook author appropriates community property for his personal profit.

III

TEAMWORK

Norms of Independence and Individualism

The recipient of a gift has no formal right to it, and the giver is not formally obliged to make the presentation. Hence, gifts are felt to represent positive sentiments rather than contractual obligations. In everyday life, efforts are made to surprise the recipient with the gift, the gift being something he is not expected to expect, and in polite society great efforts are made to select gifts with purely ceremonial rather than utilitarian value. The ideal gift is completely useless but priceless, priceless for the donor and the recipient alone. An object of conspicuous consumption often serves as second best.

When the donor is required to contribute the item, or the recipient to accept it, and when the nature of the item is specified, contractual or quasi-contractual exchange takes place, not gift exchange. Thus, a system in which gifts of information are exchanged for freely awarded recognition cannot exist in science unless scientists are free to give and receive unspecified items. The theory of scientific organization presented here implies norms of independence.

Norms of independence are very strongly held in science. Three of them can be distinguished. First, the scientist is expected to be able to select research problems freely. Second, he is expected to be able to select freely the methods and techniques to be applied to them. Third, he is expected to be free to evaluate results, to decide himself whether his results and those of others are valid or invalid. Scientists tend to take for granted the latter two norms, which they share with other professionals, and when scientists refer to "freedom in research" they usually mean freedom with regard to problem selection.[1]

The most convincing evidence for the degree of conformity to such norms would come from detailed studies of the ways in which scientists make decisions. Procedures differ among the various research establishments. In universities, norms of independence for scientists overlap norms of academic freedom. In organizations other than universities that conduct basic research, supervision tends to be highly permissive. Many give basic researchers almost complete freedom in the choice of problems.[2] Scientists who become administrators and acquire the formal authority to direct the research of subordinates tend to exercise that authority reluctantly. This is true in university research institutes as well. One of my informants was a Nobel-Prize-winning physicist who headed a large research institute in physics. He said he did not propose experiments for others to do, although he criticized their experimental designs and helped interpret their work. Since becoming an administrator, he had ceased doing experimental work himself, partly because he felt he couldn't be fair to others if he had his own research plans.

Not only formal authority in organizations, but informal "leadership" in science might be expected to compromise the norms of independence. For example, it would be reasonable to expect those who have made great discoveries to attempt to influence the work of others, and it would be reasonable to expect others to look to them for direction. However, I observed no situation in which this clearly occurred. Leaders themselves denied it. For example, a Nobel-Prize winner had a number of his former students among his departmental colleagues. He asserted, "I've left them to make their own decisions. I don't think there should be such a thing as a director of research, even at the graduate student level." Another Nobel-Prize-winning physicist was pressed on this point with the question whether he had no convictions that he tried to impress on others about what research should be done. "Sure I have convictions. But it is hard to get others to accept them. If they don't want to do it there is nothing I can do." Such statements may reflect undue humility, or they may be an attempt to impress the interviewer; even so, they express the norm of independence in problem selection. The strongest statement on this subject was made by an experimental biologist who had a distinguished research reputation and was active in scientific societies. He was asked how scientific leaders lead, and he responded, "There you have a strong norm. Telling someone what to do is *taboo*. The greatest man in science cannot tell the lowest what to do."

These norms usually apply to professional scientists only. Laymen are beyond the pale, students are defined as incompetent to make such decisions, and technicians are paid to follow directions. However, scientists tend to apply the norms of independence even to such persons. Although most scientists play a role in selecting problems for students, for example, some believe students should be required to select their own. (This will be discussed at greater length elsewhere.) A mathematical statistician said:

> My policy is not to suggest a thesis problem. . . . One of the things a Ph.D. has to show is the ability to find problems. If he can't do it for his dissertation, who is going to help him find problems once he gets his degree?

This professor had no students when he was interviewed, largely because of this policy. Finally, even as some scientists expect students to be independent, others expect research technicians to be imaginative and will respond to their suggestions about the direction of research.

In addition to studying the ways in which decisions are actually made, the degree of conformity with norms of independence could be ascertained by asking scientists about their own values. Some public-opinion surveys have done this, but their results are not especially useful.[3] West showed that there was no statistically significant correlation between commitment to the norms he studied and productivity (as measured by publications); in fact, the less productive scientists tended to be more strongly committed, although the differences were not statistically significant. He concluded, "It seems undesirable, therefore, to regard the classical morality of science as more than fortuitously associated with productive research."[4] It would be a mistake, however, to confuse the functions of commitment to the norms for *individual* productivity with their functions for the productivity of the *scientific community*. There are very good reasons for believing that conformity with norms of independence is necessary for the functioning of science as we know it. Let us consider these in some detail.

FUNCTIONS OF NORMS OF INDEPENDENCE

Workers are usually given autonomy when, during the course of work, they acquire necessary information about the job that others cannot obtain without great effort. Conversely, workers can be closely supervised only when superordinates can easily acquire information about the tasks assigned. Thus, tailors are more likely to acquire autonomy than garment workers, and personal stenographers more than members of a stenographic pool.

A professional worker acquires vital information about his task that is difficult or impossible to transmit to others. Clients are not capable of obtaining and evaluating necessary information themselves, and, even when they can obtain the information, they are less likely than the professional to act wisely on the basis of it. A physician, for example, learns more about his clients' bodies than they themselves know, and a physician would usually be unable to present this information to his lay clients in a form they could understand.

Much the same is true of scientific work, both pure and applied. The problems of a scientist may be understood fully only by a fraction of the professionally trained persons in his own discipline, and even among them only a very few will possess exactly the information he does. As a result, scientists are almost necessarily given freedom to select and apply various techniques for solving the research problems in which they are engaged.

Two qualifications must be made on this point, however. First, workers often give the impression that their tasks are highly complex and almost incommunicable to others, when, in fact, the tasks may be simple and easy to understand. Workers behave this way to maximize their autonomy, something most workers value, and to increase their power over clients and others. The automobile mechanic can control the pace of his own work, invest his job with originality, and charge outrageous fees, when he can convince his clients that his tasks are too complicated for them to understand. Physicians may do the same and, in addition, are able to induce their patients to comply with their orders and engage in necessary therapy when patients see themselves as incompetent to make decisions. And the scientist may perform technically unnecessary "snow jobs" on others to maintain his individual autonomy and prestige. Most professions involve a mystique supporting the claim that professional autonomy is necessary if a good job is to be done.

Second, because autonomy of this sort depends only on the special access of the worker to necessary information about the job, it can be limited if the superordinate is willing to make the effort and undertake the expense to obtain the information himself. Sometimes this can be done by such simple procedures as requiring workers to make lengthy reports on their work as it proceeds. Or supervisors may be recruited from among highly skilled workers and closely follow work as it is in progress. Laymen are seldom able to exercise this kind of control, but organizations are sometimes able to do so over professionals as well as other workers. In some situations, the greater control, and the increases in efficiency stemming from this, are felt to outweigh the costs of gathering the information. This is sometimes true in applied research and often true in industrial development. Thus, autonomy in the course of work, while often necessary, may be abridged.

Basic science is unlike other professions in that its practitioners not only claim autonomy in determining procedures to be used in the course of the work and in evaluating the success of these procedures; they also claim the right to decide for themselves the problems they should select and, on the basis of their work and that of others, whether or not theories are true. Other professionals may sometimes claim such rights, but these rights are less integral to their tasks than to those in basic science. For example, problems of illness are given to the physician, and, while physicians specify the problems of patients in their diagnoses, the functional specificity of the physician's role means that he is not expected to give treatment or advice with regard to other problems. Similarly, physicians depend on medical science for their theories of illness and health. The problems of lawyers are also given to them by their clients, and the validity of the principles put forth by lawyers is decided, in the last analysis, by governmental authorities—by courts and legislative bodies. In basic science, on the other hand, the selection of problems and the evaluation of the validity of theories are essential parts of the professional task.

Problems are not "given" to basic scientists by others or by "nature," at least in the most important instances. Rather, problems are discovered and invented by scientists. Sometimes problems are "obvious," such as, for example, those resulting from the discovery of paradoxes in set theory around 1900, or the problem of determining the law of electrostatic forces in the eighteenth century. At other times the discovery of the problem is the central aspect of the scientific task. Examples are numerous: it was not obvious in the early nineteenth century that the distribution of moraines and other evidences of glaciation would become problems for geologists; it was not obvious before Darwin that the distribution of species was a problem for biologists; few physicists before Einstein considered the grounds for assuming absolute space and time to be a valid scientific problem; before the work of Bolyai and Lobachevsky, few mathematicians considered the construction of non-Euclidean geometries an interesting problem. In these and many other cases, theoretical innovations and the problems they were designed to solve came, in large part, at the same time. The discovery that aspects of nature are problematical is often a central part of the discovery of the solution, and scientists may receive credit for discovering problems even if their own attempts at solution are faulty. In the long run, then, the scientific enterprise depends on the freedom of scientists to select the problems on which they will work.

In the short run, however, this need not be so, and scientists may be highly productive even if their choice of problems is restricted. Many, perhaps most, basic scientists have their freedom in this respect restricted at least to some degree. Basic research in industry is often called "exploratory" research or "background" research, and in such research scientists are free to select problems within delimited areas of interest to the firms employing them. Industrial chemists, for example, may be expected to do research on hydrocarbons but may be given great freedom of choice within this area. In universities, scientists are given almost complete freedom of choice but may be expected to do research in the discipline represented by their department. Although most of the time such restrictions are self-imposed or readily accepted, the restrictions are nevertheless present. Because it appears that such restrictions are compatible with good scientific work, they are more easily imposed than they might otherwise be, and they thus present a particularly insidious form of the subversion of scientific norms. For, in the long run, new problems are often perceived to be in the jurisdiction of no established discipline, or problems are more readily discovered by scientists outside the discipline in which they apparently fall, and restrictions on the freedom of choice of scientists inhibit basic scientific advances. This problem will be considered at greater length in Chapter IV.

The functions of the norms that scientists ought to be free to select their own problems and free to select methods and techniques have now been discussed. The remaining norm of independence, that scientists ought to be free to judge the validity of their results and the results of others, is often taken for granted, partly because such norms are thought to be widespread in our society, especially in science. This is not so; when actions depend on specific information, actors are unlikely to give those who obtain the information final say about its validity. For example, when industrial developments are to be undertaken, industrial managers are unlikely to defer to the opinions of engineers regarding their investigations; they will attempt to assess the evidence themselves. In science, however, the norm of independence is intimately connected with the goal of the entire enterprise. This goal is not to effect changes in matter or in social relations, but rather to *change beliefs*, and the basic character of science depends on the methods thought to be appropriate to that end. Insofar as methods other than the logical presentation of evidence are used or thought appropriate, the enterprise is no longer a scientific one.

Thus, the norm is taken for granted because of its fundamental nature. Problems selected because of their practical significance for nonscientists may lead to scientifically valid findings, and the prohibition of some techniques may induce scientists to discover other techniques; yet the proscription of some theories, or the prescription of others, directly prevents the practice of science. While scientists may adapt to pressures of this sort in various ways,[5] such adaptations involve *private* conformity to scientific norms: without such conformity the activity ceases to be scientific.[6] On the other hand, adherence to the norm that scientists should be free to evaluate theories helps maintain conformity to the norms of independence with regard to the selection of problems and techniques. Since decisions like these often depend on the validity attached to certain theories, any restriction of free choice concerning them may be felt to imply restrictions of independence with regard to the evaluation of theories. Similarly, if the scientist's choice of problems is restricted, this is usually because other goals, those of industrial firms or government agencies, are felt to be primary; nonconformity with the norm of independence means deviation from the goals of science.

Individual independence in making decisions about research programs probably contributes to the efficiency of scientific research, even in the short run. More important, it is a prerequisite of the continuing development of science and of conformity with even more basic scientific values. The norms of individual independence seem to be observed, and the ethos of science can be described as individualistic. Like other individualistic communities, the scientific community is highly competitive, as has been shown in Chapter II. Individualism not only leads to competitive behavior; it makes teamwork of all kinds more difficult. Yet scientists frequently become dependent on one another, and much research requires the joint efforts of scientists. Furthermore, individualism often induces a feeling of social isolation, a feeling many scientists experience as a deprivation. Thus the norms of independence in science, and the individualism they imply, stand in tension with the fact of interdependence and the desire to overcome isolation.

STRAINS BETWEEN INDIVIDUALISM AND INTERDEPENDENCE

Ordinarily the strains between individualism and interdependence are lessened by the exchange of gifts. The exchange not only permits both donor and recipient to retain their independence; it also ceremonially demonstrates their independence while simultaneously linking them in a solidary relationship. The scientific worker may search through the literature for solutions he cannot himself obtain and, when he finds them, credit the donors for their contributions in his published research. Or, he may directly approach a colleague for assistance and express his gratitude in acts of deference and a readiness to reciprocate.

This type of exchange is often insufficient. First, the scientist may require close and continuing co-operation from a professional colleague, when colleagues possess different kinds of skill and both kinds are required for a problem. Problems in virology may require the skills of both virologists and biophysicists, problems in astronomy may require the skills of both astronomers and physicists, and so forth. When continuing co-operation is required, or when large efforts are required from each participant even if they do not work together (as, e.g., when organic chemists prepare compounds that physical chemists study from a different perspective), the deference of the recipient, or his willingness to reciprocate, will not be enough to elicit co-operation. A group must be formed for a consciously sustained co-operative effort.

Second, although a scientist may possess the skills necessary for all parts of his task, the task may require simultaneous efforts by more than one individual. Even if such simultaneous efforts are not absolutely essential from a technical standpoint, they may so speed up the work that they prove essential from a competitive standpoint. Here deference and similar forms of recognition are inappropriate—the donor would not be contributing *his* special skills, and co-operation might be perceived as degrading. Again it is necessary to form a group of some sort.

A variety of co-operative arrangements is made to solve these problems of interdependence. First, scientists may agree to share the forthcoming recognition for a discovery; this usually means an agreement for joint authorship of publications. I shall restrict the term "collaboration" to this type of arrangement. Collaboration may be arranged on an *ad hoc* basis between freely acting individuals, or it may be built into the organization of a research institute. Second, scientists enlist the help of others, nonprofessionals, in return for something other than scientific recognition—although some recognition may be shared as well. The most common and important team of this kind comprises the university scientist and his students; students, in return for their assistance, receive training and, frequently, assistance in finding jobs. The help of others may also be enlisted by paying them; those who contribute to research for this reason will be called "technicians," whatever their professional standing happens to be. These are distinct forms of organization and must be discussed separately. "Interdisciplinary" collaboration must also be considered separately because of its distinctive characteristics.

Free Collaboration

COLLABORATION WITHOUT A DIVISION OF LABOR

When free collaboration arises from the interdependence of efforts, some sort of division of labor results. When it arises from a desire to avoid isolation, a division of labor is far less likely to occur. Such is the case in the formal sciences. Mathematicians work on a wide variety of problems, few of which are put to them by "nature" or practical exigencies, and, more than other scientists, they are likely to require social confirmation of the importance of their problems. They may seek this in informal conversations about their work with other mathematicians; they may also seek to induce others into a collaborative relationship. Furthermore, mathematical problem-solving is more likely to require "insight" than problem-solving in other areas, and difficulties are likely to arise because of psychological sets. Sometimes such difficulties are overcome readily by others who do not share the particular psychological set. In the mathematical sciences (and here such fields as theoretical physics may be included) more than in other areas, the Greek proverb, "When two go together, one sees before the other," is especially meaningful.

It can be seen that a sharp division of labor would fail to satisfy the motives leading to collaboration in the mathematical sciences: it would neither reinforce the mathematician's confidence in the importance of his problem nor aid in overcoming blocks by having two persons work on the same problem at the same time. The mathematicians interviewed in this study confirmed this conclusion. When asked if there was a "division of labor" among collaborators, many did not even understand the question:

> There always is. *You mean you divide up the problem and work on different aspects?* No. The way I've been doing it, we both think about the thing for a time, and when that has been done we both go on to the next.

Others were more definite:

> When we are together we talk about everything we've been doing. When we are separate each works—sometimes each man works on a different question, but this is not binding. There is some division of labor, but it is not sharp.

When there is a division of labor, it may pertain only to the trivial aspects of the work, or it may follow from a period of intimate contact and close confidence in one another. A mathematical logician who collaborated a great deal felt there was often a division of labor:

> Especially with the horrible task of writing it down. Sometimes different authors write different sections, but this is only possible when you are in sympathy with the other, or you won't like what he does. Often, when an older man collaborates with a younger man, the latter makes up the bibliography. And sometimes one person writes the first draft and another the second.

Again, a division of labor, if it exists, may develop only after work has begun: mathematicians do not often seek out another person because of his specific expertise:

> *Is there usually a division of labor in collaborative efforts?* I feel strongly that there is no point in publishing jointly unless there has been an equal contribution by both sides. *I mean contributions in different parts of the paper.* Yes, this is usually the case. I have usually done most on one part, the other on another. *Is this usually cut out in advance?* No.

When another person is approached because of his special skills, the relationship may be temporary and may take something of the form of a student–teacher relationship. One mathematician who had recently joined the institution where he was interviewed described such a case:

> One reason I came here was that X and I had common interests. Y, who is also on the faculty, is also interested in ordinary differential equations, and we have some common interests. We have done a little bit of work together as consultants at [a missile development firm]. . . . We solved one

problem out there as a joint effort, which appeared only as a memorandum. This problem was to enable me to learn something about a technique that he's expert in, more than anything else. . . .

What has been said about mathematicians in this respect also applies in large part to logicians, statisticians, and theoretical physicists. There are differences: the problems of theoretical physicists and statisticians are often given by nature, and such scientists often work closely with experimental scientists. But, when they collaborate with others in the same specialty, a sharp division of labor is unlikely.

Collaboration of this type usually begins informally; it often arises as the result of informal conversations in which two mathematicians discover shared interests in a problem. In fact, the difference between collaboration of this sort and the informal contacts described in Chapter I is very small. A mathematician may acknowledge the suggestions he received from others in his published papers, or he may insist that they jointly write the paper if he feels their contributions are important enough; the criteria of "important enough" are inherently vague.

When collaboration is initiated this way, possible partners may approach it very gingerly, even as boys and girls do not, at first meeting, suggest the possibility of romantic collaboration, although this may be very much on their minds. There is often a confession of need and inferiority: "I can't solve this problem: can you help me?"—a confession of dependency that may be abused by the person addressed. Requests for collaboration may be rejected, and such rejection is taken as a personal affront. Thus the informal communication between potential collaborators often takes the form of successively greater commitments to cooperate, much as communication in courtship takes this form.[7] And, as in courtship, the fear of being rejected or exploited may be a powerful barrier against collaboration. It is possible that scientists usually approach those of roughly the same levels of skill because of this. An inferior will hesitate to approach a superior for fear of being rejected, and a superior will hesitate to approach an inferior because of the small likelihood of receiving significant assistance. This possibility was raised in a number of interviews. Most scientists did not report patterns of the type suggested, possibly because of a rough homogeneity of skill within particular departments, possibly because differences in skill are compatible with collaboration if there is a division of labor. However, one or two felt there was such a tendency. An algebraist who did about one-third of his work in collaboration with others said that he discussed his work with departmental colleagues:

as much as I can, as much as I can find people who are interested in specific things. *How many can you find right now?* At least five in this department. In fact, that's how most of my collaboration gets started. I think of a problem and can't get anywhere and get together with someone else. *Might not*

> *this situation be psychologically difficult, when one man confesses his inability to solve a problem to another?* I think we all know that it's hard to solve problems. We know that just talking it over is often of great help even if an answer isn't gotten immediately. Just by talking about it you help yourself. *Do you suppose collaboration is most likely to occur between people with roughly equal skills?* I think that's usually the case.

When equals collaborate, one partner is unlikely to make overt attempts to direct the efforts of the other, so that the norms of independence are likely to be observed; tendencies toward authoritative direction are more likely when people of unequal skill collaborate.

In addition to arising out of informal conversations, collaboration may come about as a more direct response to the possibility of competition. When a scientist finds out that both he and another have partial results in the same problem area, he may seek to induce the other to collaborate rather than attempt to compete. While this type of information is most likely to be obtained through a scientific grapevine, it may also be obtained through publications. An algebraist thought this was the way in which collaboration usually came about:

> It arises from noticing that results someone has published have some contact with your own. If so, you send your results to the other man and inquire about the possibilities of collaboration. . . . You will know him professionally, having read his work and met him at meetings, but you may not know him very well personally.

A mathematical logician who collaborated a great deal thought casual contacts were of greater importance:

> [Collaboration] is more likely to come out of face-to-face contact, but it also frequently happens that two people will have partial results in the same specific area. When they find out about it, they agree to collaborate.

It is clear that when two or more scientists are working on the same problems they must either collaborate or compete.

A collaborative relationship, once begun, may continue for a long period. Of course, if one partner contributes very little, it will not. There are presumably scientific parasites, men who build up a large list of publications on the basis of work done by collaborators. One informant, when he first arrived at his position on a mathematics faculty, was approached by a very personable colleague with the suggestion that they collaborate on a certain problem. The informant was rather contented with such an arrangement until he was told by others that his potential collaborator had done the same many times before, that his skills were not adequate to make significant mathematical contributions, and that, therefore, he tended to exploit his collaborators. On the other hand, if both partners make roughly equal contributions, the collaboration may be repeated. One informant published almost all his work with another member of his department. The two met as graduate students and have worked together ever since. They meet informally several times a week to discuss their work, often at dinners and in other informal settings. There is no division of labor. The informant felt that this collaboration made him more effective, partly, he said, because "if you have a vague idea and are forced to explain it, in the process you are forced to make it precise, which is what a mathematician is supposed to do." Of course there is the danger in such close collaboration that possibly profitable ideas might be dropped if opened to criticism at an early stage. This was suggested to the informant, who disagreed:

> I don't think so, I think it's quite the opposite. If collaboration works well, two brains will be working with the same idea and you can tell quickly if it should be dropped. Of course, this takes a sympathetic collaborator. We just happened to hit it off, and it works well.

This is clearly an unusual case: usually such close collaboration will be avoided because it restricts the individual scientist's freedom of choice. Nevertheless, there are many cases of collaboration, which, if not as close as that described, continue for many years on a variety of different problems.

FREE COLLABORATION INVOLVING A DIVISION OF LABOR

Especially in experimental sciences, research on a particular problem may involve a wider range of skills than a single individual can possess. The scientist working on such a problem must seek to involve others in it, and this may involve collaboration. For example, a molecular biologist pointed out that some of his research required the use of complex equipment with which he was unfamiliar:

> There is quite a bit of teamwork in our department. Sometimes it leads to joint publication, at other times the contributions are listed in the acknowledgements. . . . For example, I might want to collaborate with X, since he is the only one of us to do physical chemistry with the ultracentrifuge. *Doesn't collaboration of this sort more or less reduce one of the participants to the role of technician?* Most of them are too strong to allow that. We will do minor collaborations more or less as a service. If collaboration is more extensive, it becomes more or less settled after a while that the work will end up as a joint paper. . . . *I suppose there is the problem of interesting the other man, whose assistance you need, in the merits of your problem.* You talk about the possibilities of the research to the other. He may respond with enthusiasm and then the thing snowballs. If he sees snags in it, this sort of cools the project off—you go back and think about it.

As this quotation indicates, technical services may be exchanged for rights of reciprocity and deference, but if enough work is involved these rewards will be deemed inadequate and will be supplemented by a share in the recognition for the finally published research. In any case, the donor of the services must first be interested in the possibilities of the research. This was further indicated by a biochemist:

> We don't have an electron microscope, but there is one in Biophysics. A person over there took pictures of our preparations. Similarly, there is an expert in X-ray diffraction in biophysics over there. *Is this merely using the techniques of another, or does it involve intellectual collaboration?* We want to get together to pool ideas, talk things over. . . . *Do you have any problems in getting access to others, in persuading them to take an interest in your problems?* There are always some problems like this. You have a problem, maybe the other person has different problems. You try to arouse his interest in your problem.

This shows that the professional scientist typically resists any tendency to be degraded to a technician. Either his contribution is small and exchanged for deference and rights to reciprocity, or it is large, in which case he shares in the professional recognition of the results. If a scientist wishes to have an expert in such techniques as X-ray diffraction at his disposal, he must hire a technician; the professional scientist is defined as one who freely contributes his services.

Although collaborators in this type of joint work may share in the recognition given the published papers, they may make the division of labor altogether clear. Usually the scientist who initiates the problem will sign his name first, assuming major responsibility for the results, and the scientist whose technical assistance was sought will sign his name second. (The order of signatures is capable of being perverted, and in papers produced by scientists in complex research organizations it may represent an order of formal authority, a "pecking order" as one informant put it.) Thus, in most of the journals of experimental sciences the order of signatures is felt to represent the order of importance of contributions, whereas in most of the journals of the formal sciences collaborators assume equal responsibility for the work—there is no division of labor—and signatures are usually in alphabetical order.[8]

Because the order of signatures on jointly written papers in the laboratory sciences is felt to have some importance, scientists may obtain the co-operation of others by allowing them the position of senior author. A taxonomist who was interviewed was dependent on others for specimens and other forms of technical assistance, and he professed to use this technique:

> I'm a congenital junior author. *Do you collaborate with students?* Not normally with students. With correspondents maybe. The monograph coming out has an [alien] as a senior author. He came here with a monograph, we more or less threw it out, and I wrote one. I did all the work; he was essential because of his knowledge of [the language], that's all. But he was

> senior author. . . . I also have a long association with a [person in a different biological discipline]. In that case there is a true division of labor. Actually, the theory is mine and I write the papers; he is sort of a technical assistant. I have found that if you want loyal collaboration, you can do it by making others the senior author. In the long run you get enough recognition. Others may not do it this way.

In this case the exchange of information for recognition is stated explicitly and rather cynically by the informant. Clearly, there are limits to this practice. In any case, scientists elicit co-operation from their formal peers by offering to share recognition from the scientific community for published work.

The preceding quotations have suggested that collaboration involving a division of labor is entered into more directly than collaboration that does not. A scientist who is an expert in some rare technique is expected to help others, and others confess no general ineptitude in asking him for specialized assistance. He may refuse to provide as much assistance as is asked for, on the basis of limitations on his time, and if this occurs those who seek his assistance will have few inhibitions about presenting a sales talk contrived to build up his interest and enthusiasm. Because collaboration involving a division of labor is less likely to require a marriage of minds than collaboration without a division of labor, the subtleties of courting behavior are less likely to be displayed in the former than in the latter.

The reticence of scientists to confess their inability to solve problems is unlikely to present a barrier to collaboration when a division of labor is involved. However, the sciences in which a division of labor is most likely to arise possess other characteristics that inhibit collaboration. While formal scientists may, and do, collaborate as individuals, experimental scientists often work in teams including students and postdoctoral fellows; as a result, collaboration between scientists often involves collaboration between *groups*. This means two different forms of organization must be combined, the free collaboration of formal peers and the hierarchical organization of faculty–student groups. This provides a barrier to collaboration since ambiguities in authority relations are generated. The give and take among collaborating members of the faculty may weaken the authority of each over his students, and obligations to students may not be consistent with obligations to collaborators. For example, if desirable work is assigned to the student of one faculty member it will deprive his collaborator's students of the chance to do it; working with students may be slow and put a strain on collaborating partners; and deciding whether students should join in the preparation of manuscripts will become a problem.

A theoretical physicist suggested this possibility when attempting to account for the fact that he did not collaborate with his departmental colleagues, although he wished to. This scientist had a rather large group of students (seven) working on dissertations, and he also had a number of postdoctoral fellows working with him. In the past he had collaborated with departmental colleagues but was not now doing so:

> Some members of the faculty would work in this [problem] area if I weren't already, and they don't know if it would be appropriate or how to go about it. I think this is somewhat detrimental to general progress. I don't know what to do about it. For example, the man who shares this office with me had done work in this general area; he is a junior man and had done work in other areas and now seems to have steered away from this area. He now works on problems which I think are not his first choice. It is hard to know why. . . . *Perhaps to avoid competition between the two of you?* There is none between us—we are far apart in age and rank. . . . It is the academic tradition that individual faculty members should be independent. This has broken down in experimental physics here, but continues to be the tradition in theoretical physics in a single department. The exceptions occur when the tradition is broken right when a person enters the department. For example, the man of whom I've spoken has formed such an association with another assistant professor, and they have worked together very effectively. I did the same thing many years ago at X university with Y, who arrived there two or three years later. . . . It is complicated, and I can't understand where it comes from. Maybe it's the old business of empire building. In many areas of science, when a senior faculty member has accumulated a group of junior persons, it becomes difficult for him to collaborate with another senior person; he gets identified with his group. Thus, for close collaboration between two faculty members to occur, it must begin before one of them has formed a group.

As the informant indicated, when one man has not formed a group it is easier for him to collaborate with another. He himself had collaborated very successfully, about two years prior to the interview, with a visiting professor who was not associated with a group of students and postdoctoral fellows.

When two experimental scientists both of whom are associated with groups collaborate, they may attempt to keep their groups out of the relation. For example, an eminent molecular biologist had a large group working with him—five graduate students, two fellows, and four technicians. He had entered into collaborative work with another eminent man at a university a few hundred miles away:

> We exchange ideas and we have done some experiments together up at X university. I brought some materials back here to analyze. If this goes on long enough I suppose we'll have a joint paper.

But this does not involve co-operation between the respective groups:

> This is really an individual operation. . . . Members of my group may do some analyses. Technicians in the group carry out experiments for me. I

don't like to have fellows or students do this kind of work—they have their own problems.

In an experimental physics division at one university, the problem was partially resolved by placing collaborative relationships among the faculty on a higher level than relations between particular professors and graduate students. A student was not formally assigned a single professor as his adviser or research director, but, instead, reported to all members of the staff. A member of the staff reported that the faculty divided up the students:

> in a rather informal way. I have six and one-half of them—who report more to me than to the others. . . . They choose [the dissertation topic] themselves in consultation with others of the staff. We try to avoid having the boys talk to one staff member only. And we have frequent changes—the list of assignments is kept away from the boys . . . In a lab of this kind there is much sharing of equipment—there is a limited time on the accelerator, so the whole thing must be definitely a co-operative effort. No one is a formal boss. All the full professors share equally.

The university can arrange this only by dividing up its physics staff into about four relatively independent groups, so that the one referred to could have but four men of professorial rank and twenty-eight graduate students. Professors in the different groups may find it difficult to collaborate with one another, for the same reason professors elsewhere who have groups of their own.

Despite these procedures for combining student teams with faculty teams, there is a negative correlation between them. The data presented in Table 11 show that University of California scientists with many graduate student research assistants were less likely to say they worked in groups with other members of the faculty than those with few graduate student assistants. Those with no students were least likely to collaborate with peers, but this category includes newcomers to the University and men with few research interests.[9] The correlation between the two variables is not especially large (in the first row of the table, nineteen percentage points separate those with one or two students from those with five or more), and its size is probably reduced because both variables are confounded with research abilities and "sociability." Men who are highly productive and gregarious can be expected to attract more co-workers, students as well as faculty. In my sample there was also a low correlation between two related variables. The correlation (Q) between number of students and a crude measure of the extent of intradepartmental communication was -0.01, whereas most of the other correlations between various types of communication practices were high and positive (see Chapter I, Table 4).

The difficulty of combining hierarchical organization within groups

TABLE 11

Collaboration with Students and Peers—University of California Scientists*

NUMBER OF OTHER FACULTY MEMBERS IN RESEARCH GROUP	NUMBER OF GRADUATE STUDENT RESEARCH ASSISTANTS**			
	NONE	1 OR 2	3 OR 4	5 OR MORE
None	85%	56%	70%	75%
1	5	19	10	11
2 or more	10	25	20	14
TOTAL	100	100	100	100
(Number of cases)	(40)	(36)	(30)	(28)

Source: Secondary analysis of data collected by Robert Wenkert and Roderic Fredrickson.

* Formal, physical, and biological scientists who are members of the faculty and are engaged in research, with the exception of physicists, who are excluded.

** The question asks about the number of "research workers who do not have a Ph.D. or its equivalent," but in context this almost certainly means graduate student research assistants. These assistants were counted in terms of half-time equivalents.

with free collaboration between peers makes it likely that free collaboration occurs more often among formal scientists, who do not require groups of assistants, than among experimental scientists. This has been suggested by Norbert Wiener:

> This habit of joint work is almost the peculiar property of mathematicians and mathematical physicists. Most other scientists are hampered by the fact that they are dependent not only on a laboratory but on a very special laboratory with their own materials and equipment.[10]

This does not mean that mathematicians are more likely than other scientists to write papers together. Scientists in other disciplines *must* work in teams, but the teams need not involve free collaboration. Disciplinary differences in collaboration will again be considered farther on in this chapter. First, however, the ways in which scientists select collaborators will be examined in greater detail.

THE CHOICE OF COLLABORATORS

It has been suggested that collaboration involving no division of labor is likely to occur between individuals of roughly equal abilities. Other factors also affect the choice of collaborators. Since collaboration often begins through informal contacts, anything that increases the frequency of such contacts increases the likelihood of collaboration. In the present study, scientists who had more informal contacts with their departmental colleagues were more likely to collaborate in joint efforts. This is shown in Table 12.

Spatial propinquity often leads to collaboration since it is likely to lead to informal communication. This is not to say that face-to-face contacts

TABLE 12

COLLABORATION AND INFORMAL COMMUNICATION*

PROPORTION OF COLLABORATIVE WORK	RELATIVE AMOUNT OF INFORMAL COMMUNICATION WITH DEPARTMENTAL COLLEAGUES	
	High	Medium and Low
High	54%	21%
Medium	27	26
Low	19	53
TOTAL	100	100
(Number of Cases)	(26)	(34)

* See Ch. I, appendix, for details on classification and measurement.

are *necessary* for such collaboration, for it can be and often is carried on by letters at great distances. But, since scientists usually agree to collaborate after discussing research informally, an absence of informal contacts makes collaboration far less likely. This was noticed by some scientists after changes had occurred. In one case, after half of a physics department had moved into a newly completed building, one of the others reported, "We see people who have moved there much less frequently than before," and chances for collaboration were reduced.[11] Again, a physical chemist reported that he had fewer contacts with specialists in the theory of the solid state in his new institution than in one where he had worked earlier:

> That is done mostly here in the electrical engineering department, and I don't know too many electrical engineers. It's a question of proximity; they're in a building five minutes' walk away. At X university they were in the next building and I had lots of contact with them. . . .

On the other hand, when collaborative relations involve a division of labor, scientists are likely to seek out experts even if they must travel some distance.

Caplow and McGee suggest that the calculation of prospective gain or loss of prestige is an important factor in the choice of collaborators. Taking formal rank as the measure of prestige, they suggest:

> . . . [We] would expect that collaboration with peers would be avoided, since it inevitably involves the participants in each other's reputation, whereas the prestige system holds them in competition with one another. . . .

> The junior members of the department may feel [should a man of lower rank collaborate with a full professor] that the man in question is getting an inside track by playing up to his senior. The other full professors may feel that their colleague is threatening their collective status by working too closely with an inferior. . . . The professor will eventually have to help decide on the promotion or nonpromotion of the junior man. Will their close association be a source of prejudice? . . .
>
> We would expect, and we find, that full professors most frequently collaborate with others of their own rank. Collaboration between ranks, when it occurs, takes place between adjacent ranks, since the status distance and the threat of status compromise are less between adjacent ranks.[12]

Caplow and McGee's findings are necessarily tentative, since they had data on the relative ranks of collaborating individuals in only fifty-one cases; moreover, tendencies of this sort could be explained by the greater chance that full professors engage in informal interaction with one another. Full professors will have had time to build up friendships and will have less social distance separating them. It is not necessary that conscious considerations of possible status gain or loss enter into the choice of collaborators.

The scientists interviewed for this study failed to see the pattern suggested by Caplow and McGee. While they may have been reluctant to reveal such status-striving considerations, their answers suggested that propinquity was the most important factor. In answer to the question, "Are you more likely to collaborate with persons of some ranks and ages than others?" a theoretical physicist replied, "Most people I know are older, so my collaborators usually are." An assistant professor in mathematical logic said he did not find it difficult to collaborate with older men, although:

> Older people may have more responsibilities and thus have less time for idle talk, and for this reason they may be less likely to start collaboration. But they also know more problems and are more likely to make suggestions.

Both spatial propinquity and social distance are important, for essentially the same reason. When the social distance between ranks is large, one can expect collaboration between individuals of unequal ranks to occur less often than collaboration between peers.

Finally, when collaboration emerges gradually from informal contacts, any such "irrelevant" criteria as race, sex, or appearance that influence the probability of informal conversations will influence the choice of collaborators. Such criteria are not legitimate; the norms of science excluding personal qualities extend even to the choice of collaborators, and these norms seem usually to be followed. In an article about the work of the physicist Samuel Goudsmit, we get a picture of normal behavior:

> As he sees it, achievement arouses affection in a physicist, and he cites as an example his own feeling toward a notoriously bad-mannered but exceed-

ingly brilliant scientist. "He's a mean, caustic, and boorish man." Goudsmit says . . . "To most people, he would be *persona non grata*. To me, he is a man who has solved difficult scientific problems, and in my home, he is welcome."[13]

The scientists who were interviewed expressed similar sentiments. All the same, a study of the experiences of scientists whose ability to interact freely is hampered by such characteristics as being Negro or female might reveal that they are less likely to enter into collaborative associations, especially when no division of labor is involved. Such social disabilities can be expected to be less important when a division of labor is involved and one or both partners can be approached as an expert with a unique competence.

Teams Including Students and Technicians

The ways by which scientists obtain information determine their needs for technical assistance. It is convenient to distinguish three types of sciences according to the ways in which information is obtained. There are, first, the theoretical sciences, in which information can be obtained by an individual working alone or in a library. The theoretical sciences include not only the formal sciences but also theoretical physics, theoretical physical chemistry, sociological theory, etc. Second are the laboratory sciences, which use more or less complex and stationary instruments for gathering information. The instruments may be rather simple, so that, as in the case of many organic chemists, the scientist may construct most of them himself, or they may be as complex as the particle accelerators used by experimental physicists. Laboratory sciences are not necessarily experimental sciences, for astronomers and seismologists also depend on stationary instruments. Because their instruments are often located far from central research establishments, however, they also tend to fall into the third type, field sciences. In the field sciences, scientists engage in surveys or expeditions that take them away from research establishments; examples are natural history, geology, oceanography, and ethnography.

This classification cuts across disciplinary lines. For example, physicists may engage in purely theoretical work, or they may participate in expeditions (as in the study of auroras), although physics is usually thought of as a laboratory science; geologists may perform laboratory experiments in addition to field research; and astronomers may spend most of their time analyzing data collected by others rather than making observations themselves. The present study is concerned largely with theoretical sciences and the laboratory sciences closely related to them. Field sciences will not be discussed except in passing.[14]

In the laboratory sciences, there is usually a sequence of activities involving some preliminary theoretical work, detailed planning of the experiment, constructing or otherwise obtaining equipment, routine analyses of data, solving unanticipated problems, and analyzing experimental results. If the laboratory scientist cannot solve problems himself, he either consults another and possibly collaborates with him, purchases equipment, or hires skilled technicians to assist him. Even when he can solve problems himself, he may need assistance. The construction of equipment, performance of routine analyses, and some parts of the final analysis of experimental data can be performed by less highly trained personnel under his supervision.

In U.S. universities, these assistants are usually graduate students, while postdoctoral fellows are another important source of assistance. For example, among the experimental biologists interviewed, the ratio of postdoctoral fellows to graduate students in research groups was at least one to three. Berelson has shown that in some fields a postdoctoral fellowship has become almost a necessary step in a successful professional career.

> A leading physiologist, speaking of the "explosion" in his subject, said that post-doctoral programs would grow because the nature of the field was forcing change in that direction. A leading mathematician also used the "explosion" figure, adding that in his field the student must master the earlier material as well as the new, and concluded that "the Ph.D. has, in effect, a general education in mathematics; he needs more specialization and that is where the post-doctoral program comes in."[15]

Postdoctoral fellowships may involve independent work, but, in the laboratory sciences especially, they often consist of a prolongation of graduate work. The postdoctoral fellows in the laboratory of a professional scientist contribute to the professional's research, and they write many of his papers with him.

A final source of technical assistance is the paid technician. These range from unskilled bottle-washers and animal-tenders to college graduates who may contribute greatly to research and appear as joint authors with their employers.

In laboratory sciences in U.S. universities, the assistance of graduate students and postdoctoral fellows is indispensable to a professor's research. A large proportion of the articles in such periodicals as the *Journal of the American Chemical Society* are jointly written by a professor and his students. Thus, an experimental physicist could answer the question, "Does having students contribute to your own research?" by saying "Yes—they do it," without being too facetious. A graduate dean who was a chemist told Berelson:

> [I]n the case of many a professor in distinguished institutions, the major scholarly work of the professor is simply the summation of the original work which graduate students have done in collaboration with him. Only

occasionally does a professor of chemistry write an article which is not in collaboration with a graduate student.[16]

An organic chemist who was seeking employment considered the possibility of getting good students a prime criterion in his choice of institutions:

> [The possibility of getting equipment] is definitely an important criterion in selecting a place to work. However, nowadays there is enough outside aid available if you're good and self-confident. The universities where money is a problem also have other problems—they are likely to have a poor staff and cannot attract good graduate students, and so forth. I suppose a good man in such a place might succeed eventually in attracting good students, but it's much better if the students are already there.

Laboratory scientists may need only a few students if their work is in a planning stage, and thus unlikely to lead to quick dissertations, or if their work involves technical problems with which students are unable to cope. A physical chemist who used unusually complex and expensive equipment said, "You can handle a large number of students if you have a definite program. . . . I can't get equipment fast enough to keep them busy. . . . One year I had five and almost went insane." Most laboratory scientists, however, desire to get past the planning stage quickly and develop a "definite program"; doing so increases their productivity and the recognition they may receive.

Theoretical scientists are not dependent on technical assistance for the completion of research to the same degree as are laboratory scientists; if they value graduate students at all, they value them for other reasons. A typical statement was given by a mathematician who was supervising no doctoral dissertations at the time he was interviewed and was not sure he wanted to have this responsibility:

> They take a lot of time, but working with them can be very rewarding. By and large we have very good graduate students here, so it is by no means a chore in the sense that you have to keep nagging at the boys to get them to work. It isn't the activity that appeals to me most. . . . *Does it contribute to your own research?* Only in an indirect way. You may get interested in some problem because of it.

Mathematicians often see the solutions to doctoral problems long before their students and must consciously refrain from giving them too much assistance:

> It is not easy to think of good research ideas; it is a tricky problem. Problems either don't have enough meat on them or they are very important and meaty, in which case you might want to work on them yourself. . . . And you want to help the student, but not too much.

Formal scientists may value having students primarily because it gives them the opportunity to present materials in seminar discussions, not because they receive research assistance. When formal scientists were asked if they would like to have more students, only five out of twenty replied "yes," and these five included four assistant professors who had no students at all.

Apparent exceptions serve to confirm the foregoing reasoning. A mathematician who had five graduate students and thought he might want more worked in a branch of numerical analysis involving the use of computers. Essentially his work required "experimental" work with computers, and students were helpful in programming computers, running them, and getting rid of bugs. Unlike most formal scientists, much of his written work was jointly prepared with students. But such men are relatively rare in the formal sciences.

As a result of all these factors, teamwork is far more common in the laboratory sciences than in the formal sciences, although collaboration between peers involving no division of labor may be as common in the latter as the former. Data from a survey at the University of California, Berkeley, support these conclusions. Faculty members who engaged in research were asked, "About how large a research group is working with you on your own research?" The proportion in various disciplines who work in teams of any sort, as well as the composition of the teams, are shown in Table 13.

Sample biases and the wording of the question about teamwork may have increased the apparent amount of teamwork; members of groups may have been more likely to respond to the mailed questionnaire, and even very informally organized teams may have been counted. It is clear, however, that teamwork is less common among mathematicians and statisticians than among laboratory scientists. When mathematicians are members of groups, however, the groups are more likely to include faculty colleagues than groups of chemists or experimental biologists; 25 per cent of the mathematicians reported working with faculty colleagues as against 15 per cent of the chemists and 21 per cent of the experimental biologists. In the latter areas, as has been suggested, faculty–student teams inhibit collaboration between peers. Only in physics, where it was impossible to distinguish between theorists and experimenters, do groups often include students and technicians, as well as faculty colleagues.

The apparent inconsistencies in Table 13 are also interesting. When mathematicians or experimental biologists speak of working in a group, they often mean a group of faculty colleagues; only 29 per cent of the mathematicians and 62 per cent of the experimental biologists said they worked in a research group, although 42 per cent of the former and 83 per cent of the latter did have the assistance of graduate students or technicians That students "don't count" was also noted in the interviews. When asked if their published work was done in "collaboration" with others, a number of scientists replied "no," although almost all of their papers had been jointly written with students.

TABLE 13

Teamwork by Discipline—University of California

DISCIPLINE	PERCENTAGE OF SCIENTISTS WORKING WITH TEAMS					
	WITH A GROUP	WITH 1 OR MORE GRADUATE STUDENT ASSISTANTS	WITH 1 OR MORE POST-DOCTORAL FELLOWS	WITH 1 OR MORE FACULTY COLLEAGUES	WITH 1 OR MORE TECHNICIANS	(NUMBER OF CASES)
Mathematics and statistics	29	42	12	25	12	(24)
Physics*	97	93	57	50	43	(30)
Chemistry	88	88	35	15	35	
Experimental						(26)
biology**	62	83	67	21	75	(24)

Source: Secondary analysis of Wenkert-Fredrickson study.
* Includes both experimental and theoretical physicists.
** Includes those identifying themselves as biochemists, geneticists, and anatomists or physiologists.

Because of these patterns of teamwork, most of the papers in the laboratory sciences have two or more authors. The over-all results are shown in Table 14, from Berelson's study of leading scientific journals.[17]

TABLE 14
Joint Authorship by Discipline

DISCIPLINE	PERCENTAGE OF PUBLICATIONS WITH MORE THAN ONE AUTHOR, LEADING JOURNALS, 1957–1958	AVERAGE NUMBER OF PUBLICATIONS THROUGH 1955 BY RECIPIENTS OF THE PH.D. IN 1947–1958
Chemistry	83	6.1
Biology	70	5.7
Physics	67	4.9
Psychology	47	5.4
Mathematics	15	4.0
Philosophy	5	2.8
English	3	2.2
History	4	0.5

Source: Bernard Berelson, *Graduate Education in the United States* (New York: McGraw-Hill, 1960), p. 55.

In the discipline that is most purely a laboratory science, chemistry, 83 per cent of the papers are written jointly, whereas, in the discipline that is almost purely a theoretical science, mathematics, only 15 per cent are.[18]

Table 14 indicates that in disciplines in which most publications have more than one author the average number of publications per professional is higher. This is also true for individuals. Results for the present study are shown in Table 15.

TABLE 15
Productivity and Joint Authorship

	DEGREE OF JOINT AUTHORSHIP*		
PRODUCTIVITY	High	Moderate	Low
High	61%	62%	48%
Moderate	30	31	20
Low	9	6	32
TOTAL	100	99	100
(N)	(23)	(16)	(25)

* See Ch. I, appendix, for details on measurement and classification.

An even stronger association exists between the number of students and postdoctoral fellows in a professor's group and his productivity; 90 per cent of those with groups of four or more students and fellows were classified as highly productive, as against 45 per cent whose groups were smaller.[19] Those with more assistance are able to publish more, but it is also true that productive scientists are best able to attract students.

INDUCING STUDENTS TO JOIN TEAMS

Since students are an important source of research assistance, competition for them is intense in the laboratory sciences. Within a discipline, the most easily justified form of competition is that among the various universities. (Even here, however, observers from Veblen to the present have decried such competition as improper and leading to an unjustified concentration of talent in a few institutions.[20]) Departments having large undergraduate enrollments of good students are in a superior competitive position, while departments in professional schools are often at a disadvantage. The chairman of a science department in a medical school mentioned this:

> We are definitely competing for the best graduate students. *Are you at a disadvantage since you don't have an undergraduate department?* We definitely are. . . . We refrain from making a serious effort on this campus. The biochemistry division in the chemistry department is in a position to attract the chemistry majors interested in biochemistry. . . .

Most departments engage in publicity efforts to attract students. Although personal contacts at undergraduate universities are probably most important, other means are also used. One department of experimental biology participated in a television series, and a member indicated that a beneficial result of this was a number of inquiries from potential graduate students.

Within disciplines, specialties compete for students. This topic will be discussed further in Chapter IV, although it may be noted here that this is perhaps the major form of competition for students in the formal sciences. Theoretical scientists frequently compete *not* to have students. Thus, a mathematician in a leading department said, "Students compete for professors. I had to turn down three students last semester, including one very good one." A student in the same department said that the two professors in his specialty in the department "don't want students. They don't compete for them, give their students a lot of trouble, and delay them in getting degrees. . . ." When a specialty is competing for prestige within the discipline, however, its practitioners may exert themselves to obtain students—not because students can assist them on research, but because specialists feel the need to expand their area. For this reason, specialists in such marginal branches of mathematics as logic or applied mathematics may attempt to entice students away from other specialties.

Within a department, professors attract students in a variety of ways. The interest of the student in receiving an education is naturally the most important element. He may have special interests and seek out a member of the faculty who shares them. However, in the sciences students are seldom competent to make such choices of subjects; they are usually unable to select dissertation problems that are scientifically important, sufficiently easy to solve, and not too likely to involve them in competition with others. Thus, in Berelson's sample, only 2.5 per cent of the scientists said that students really select most dissertation topics, and, in physics and chemistry, a majority of the faculty said that the professor really selects most of them.[21] (In fields like sociology or history, on the other hand, less than 10 per cent of the faculty said that the professors really select most dissertation topics.) Laboratory scientists usually give their students problems that contribute to their own larger research efforts. A statement by a biochemist who was unusually successful in attracting students illustrates the procedure:

> I tell the student or the fellow that I expect him to start to work in an area in which I have some competence, and I suggest a particular problem. Just as soon as he finds it possible, and I think it desirable, he may deviate from the original objectives, and in the majority of cases that almost invariably happens. So much so sometimes that I'm sorry I allowed them to deviate so far. . . .

A less permissive biochemist put the matter more bluntly:

> You're only interested in a man if he can add to your own effort. Graduate research pretty much centers around the chairman: he knows more of the area he's working in than the student, or than anyone else, for that matter.

Usually postdoctoral fellows are more independent. They are better acquainted with their fields and select men to work with on the basis of common interests. (In addition they need not be concerned about getting a degree.) One molecular biologist said that postdoctoral fellows

> arrive here with some independent ideas, and they also want to learn our techniques and approaches. Usually the postdoctoral fellow continues with some lines of investigation he had already started, and he starts some new ones which contribute to the program I have under way. Sometimes we keep Ph.D. people on for a year after they complete their theses here. This benefits both the professor and the student—progress on the professor's research and publications for the student.

Although students may be unable to choose a professor on the basis of the substantive details of his research, they may be able to make decisions on the basis of a professor's general reputation and his productivity.

Highly productive professors are likely to have good research problems available and can offer students the ability to publish. Thus, in Berelson's sample, the more productive scientists (top 40 per cent) were more than twice as likely to have groups of three or more graduate students than the less productive (bottom 60 per cent).[22] Competition for students on the basis of productivity is often perceived to be desirable. A leader in a department of metallurgy said that, because of shortages of funds, students were not always able to work with the man they preferred, but:

> I hope to be able to give graduate students their choice of projects. Then men producing good and effective research will attract students. This is the best way of organizing things.

Of course, students are not always able to evaluate either the quantity or the quality of the research of professors; the important determinant of choice is the *perceived* productivity of the professor, and his actual productivity is only one factor influencing perceptions. A student in a high-energy physics laboratory described the way groups within the laboratory competed for students:

> There are two types of people here, high-pressure and low-pressure. The high-pressure people compete for students, in order to get work done quicker and more completely. However, they do this in a backhanded way; they don't approach particular students but rather puff the quality of their groups. They will speak about how good their groups are; they don't present their physics but their snob appeal in recruiting. It's very hard to choose a group when you're beginning--you never really know about them. You talk about different groups with your friends, very rarely with group leaders.

A professor of biochemistry felt that students' choices were often irrational, depending on rumors about which member of the faculty was "easiest" or on a wish to work together in the same group. He felt this meant some professors had more students, and others fewer, than they needed, and he thought his department should set a maximum on the number of students working with any one professor. In some departments, this type of collective action is used to regulate competition for students. While graduate students are not usually permitted to select an adviser until their second or third semester, one department more or less "rationed" students after that time:

> By the third semester the student tells the graduate adviser his preference of someone to work with. The adviser tries to distribute students equitably. If several students wanted to work with me I would get my pick. There are not quite enough good students to go around.

In this case the rationing may have been necessary because of space limitations. Usually the powerful and prestigious members of a faculty, those who have most influence on departmental decisions, will resist any form of rationing on the grounds that students should have freedom of choice and that competition is a good thing. Other factors may also be important in inducing students to join faculty-led research groups. First, it is important to the student that the professor remain with the department until he completes his dissertation, and well-established men with much seniority are more likely to attract graduate students than young men with doubtful tenure. Thus, in Berelson's sample, full professors were more than twice as likely to have groups of three or more students than were assistant professors. Second, a professor's ability to dispose of salaried assistantships or fellowships may influence the number of students he gets. At present there are many sources of money for students in the sciences, so this is unimportant in first-rate universities. The chairman of a biochemistry department said that the presence of research assistantships made "very little" difference in competition for students:

> because the U.S. Public Health Service has made money so available that just the presence or absence is no longer a factor. Anyone can get them. There may be some difficulty here in terms of the amount. . . . Today it is almost unheard of that a doctoral student would start his work on his own funds.

Berelson concluded from his national survey that "almost all doctoral students get some form of support."[23] In departments with less prestige, however, the faculty member's ability to dispose of stipends has a great deal of influence on his ability to attract students. (Such departments train a relatively small proportion of Ph.D.'s and do a relatively small proportion of university basic research.) An experimental physicist in such a department said there was competition for students but that the choice of dissertation chairmen:

> is left completely up to the students. However, the students are influenced by self-interest; the faculty member who has access to more money for financial support will attract more students. . . .

This was the only department studied in which money made a difference in competition for students. Students are apt to have the same motives as professionals; they want to perform independent work that will be recognized by others.

EXPLOITATION OF STUDENTS

The benefits of teamwork between a professor and his student ideally accrue to both, but, since the former is the more powerful person in the relation, exploitation is possible and often occurs. Berelson asked recent recipients of the Ph.D. about this:

> . . . recent recipients in the sciences complain in substantial numbers (46% agree to 34% disagree) that "major professors often exploit doctoral candidates by keeping them as research assistants too long, by subordinating their interests to departmental or the professor's interests in research programs, etc."[24]

Differences among disciplines in this regard depend on the relative importance of the student's contributions to his professor's research. In Table 16, responses to Berelson's question about exploitation are given for twelve fields. The question did not ask respondents to report about their

TABLE 16

EXPLOITATION OF GRADUATE STUDENTS AS PERCEIVED BY RECENT RECIPIENTS OF THE PH.D. IN TWELVE FIELDS

DISCIPLINE	PERCENTAGE OF RECENT RECIPIENTS AGREEING THAT "MAJOR PROFESSORS OFTEN EXPLOIT DOCTORAL CANDIDATES"	(NUMBER OF CASES)
Microbiology, biochemistry, and biophysics	57	(162)
Physics	52	(103)
Chemistry	49	(291)
Zoology	40	(84)
Geology	40	(50)
Psychology	37	(230)
Economics	33	(61)
Political Science	30	(53)
Sociology	29	(41)
Humanities	28	(229)
History	28	(93)
Mathematics and statistics	28	(64)

Source: Secondary analysis of Berelson data.

own experiences or their own disciplines, but it is likely that their judgments were based mainly on their knowledge of their own disciplines. The data indicate that exploitation is most often thought to occur in the laboratory sciences—such fields as physics, chemistry, and molecular biology. It is least likely to occur in mathematics, where, as was noted, professors are very seldom assisted in their work by students. Exploitation is also unlikely to occur in the social sciences and humanities.

The kind of exploitation referred to in Berelson's study—prolongation of graduate work and subordination of the student's educational interests to the professor's research interests—is, if common, relatively mild. A professor may also exploit students by misappropriating their work and by attempting to force them to support his particular points of view. Such exploitation has more serious consequences because the professor–student relationship is not only formal but interpersonal. An excellent example of the possible consequences is provided in Edward Lurie's biography of Louis Agassiz, the eminent nineteenth-century naturalist.[25] Agassiz was a charismatic teacher whose students often became intensely devoted to him. However, in both Switzerland and the United States, most of his leading students eventually were alienated from him. His students at Harvard publicly accused him of unrightfully appropriating their work; in addition, he seldom allowed students to pursue their own research interests. These disputes strongly affected both Agassiz and his students (for example, a disciple's betrayal helped induce Agassiz' first wife to leave him).

Agassiz was, in many ways, like a domineering father to his students. As long as they were loyal to him and accepted their subordinate status, they would be rewarded, but he would not tolerate aspirations toward independence. Other examples of such behavior are well known.[26] A few of my informants described such authoritarian relations. A formal scientist told the following anecdote:

> A student at Y university had worked with X for many years but didn't want to write a thesis on one of X's topics and so came to me. X didn't communicate about him directly to me, but through someone else. He asserted that the student was seriously disturbed and that I should not direct his thesis. Actually the student was just rather shy and happened to disagree with X. X doesn't like people who are uninterested in his ideas. He's unusual in this respect. . . . X's students tend to come back eventually to Y university or to have emotionally painful breaks with him.

The professor in question is a very eminent man, and it is quite possible that eminent men, especially those trained in European universities where professors have traditionally had more authority, tend to develop authoritarian sentiments with respect to students.

Such authoritarianism is rare. In general, one receives the impression that professors are unusually altruistic toward their students: they desire students to become independent and will sacrifice their own interests to this end, and they tolerate or even encourge disagreement on important scientific issues. (The word "unusually" is justified because, as noted in Chapter I, the self-images of professors may depend on the reactions of students, and there must be strong temptations to discourage students from adopting opposing views or developing distinctive research interests.) Conversely, graduate students often deeply respect their professors, identify with them, and use them as models for a wide variety of behavior. Professors are often selected on the basis of personal attractiveness, especially in fields where students have relatively little knowledge of the research being done. Thus, Krohn found many university scientists who idealized their advisers:

> In answer to the question, "What kind of a man was your adviser?" roughly one-fourth of all [87] university [of Minnesota] scientists praised their advisers in strong terms, both as scientists and as persons. Eight of them praised this so strongly as to amount to a vertitable eulogy. . . .[27]

The university scientists in Krohn's sample were more likely to express deep respect for their advisers than were industrial or governmental scientists, and they were also far more likely to be deeply committed to the values of basic science.

In Giovacchini's psychoanalytic study of eight laboratory scientists he observed:

> [F]antasies of a more intellectual nature involved the admiration of elderly men, revered teachers or famous scientists who were particularly fascinating to them and who recognizing the potential of the patient made him a protégé.[28]

Whereas the professional scientist receives recognition from a professional community, the graduate student at a crucial point in his career depends for recognition primarily on a single individual; his self-image as a scientist is dependent on the personal reaction of his advisers.

The dependence of students on professors, and the tendency of professors to recognize this, has consequences not only for the interpersonal relations between students and professors but for the kind of research usually performed in universities.

EFFECTS OF STUDENT TEAMS ON RESEARCH

Students are less likely to take risks in research or engage in long-term research than are professional scientists. They have a precarious status, one in which they receive few material rewards, and they therefore usually desire to obtain a degree in as short a time as possible. If an experiment fails—an hypothesis cannot be confirmed or can be neither confirmed nor confuted—the student suffers more than the professor. An experimental physicist described failures in the following way:

> Once it took a year [to find out that it failed]. Usually you have a nauseating feeling six months before, but you don't give up without a fight. You can sense it, but you aren't quite sure, and follow it to the bitter end. Although, it is not always easy to tell when an experiment fails. . . . For students, it is partly a matter of luck. If the experiment doesn't work out, they must stay on a year or so longer. It's usually because I goofed, or maybe they did.

A biochemist who was interviewed was just about to wind up work on a problem that had occupied him for three years. His failure to discover a certain enzyme meant he could not confirm a very promising theory, so his efforts essentially resulted in failure. His lack of success probably led to his having but a single student, far fewer than he would have desired.

Work with students tends to discourage risk-taking as well as long-term projects that cannot be broken up into dissertation-sized parcels. Other pressures in universities have the same effects. The most important of these is the practice of basing promotions on published research. The informant quoted above had been an associate professor for four years and an assistant professor for another four, and it is likely that a promotion was held up because of his unsuccessful risk. Risk-taking and long-term projects may also be discouraged by having to seek outside research grants; this will be discussed later. As a result of these pressures in universities, certain types of risky and long-term research are more likely to be carried out in governmental or industrial laboratories. A mathematical physicist working in a government laboratory commented on the fact that he was under little pressure to publish research:

> People in universities must limit the size of the problem they tackle so that they can publish something in a short length of time. University people probably couldn't work on my problem, since I don't expect to get results for another two years. . . . Government laboratories are freer—there is no short-term competitive problem as there is in the universities.

Probably more is involved than the external needs of students and publication pressures. Since research is done in organized groups, the status of the group leader is dependent on success in attaining group goals. Failure may mean a loss of status for the leader, or at least the leader may expect this to happen. He will therefore be tempted to avoid problems in which failure is likely.[29]

This diminishes the differences between university and nonuniversity scientists with regard to risk-taking. Industrial laboratories are usually organized in some sort of hierarchical fashion, and government laboratories often are. Groups in these laboratories are more or less permanent, unlike those in universities, so that the leader's desire to maintain or increase his prestige has an even stronger tendency to diminish risk-taking. While the mathematical physicist just quoted worked by himself in the government laboratory, a theoretical physicist in the same laboratory worked in an organized group. He described his supervisor as:

> a man under a great deal of pressure—he feels he has to get results and isn't motivated by intellectual curiosity. Unfortunately, I get the feeling that the physicists who have intellectual curiosity are a minority group.

Teamwork does not have these effects for most university research. Although students in the laboratory sciences usually work on dissertations relevant to the professor's research, each usually has a definite problem of his own, and day-to-day co-ordination with other members of the student group is not required. Only in the high-energy physics laboratories did this appear to occur. In chemistry and experimental biology, members of the group are only loosely interdependent; a professor may fail in his work with one student without jeopardizing his relations with others.

Some ways in which the dependence of professors on students limits the research they do have now been described. Evidence is not available to test other similar hypotheses. For example, professors who must compete for student assistance may avoid topics or techniques that students perceive to be unfashionable, and they may tend to avoid tasks that students cannot perform. In addition to these consequences for research, the tendency of professors to share the recognition of published work with their students may affect the way research is recognized and the meaning of recognition.

Sharing Recognition

One means whereby professors may induce students or colleagues to assist them is to offer them the chance to publish joint papers. When this is done, all those whose names appear as authors are presumably responsible for the work and share the recognition it receives. In fields where a division of labor is common, the author whose name appears first is usually given more credit. Today in the United States, professors are probably more apt to give students credit not really due them than to appropriate the work of students for themselves. A biochemist whose papers were usually written with his students or postdoctoral fellows made a typical statement:

> In my early days I had no collaborators, no research assistants, nothing; I did it all myself. The moment I got a competent research assistant or graduate student, we worked on a program, and invariably put his name on the papers, and that has been my practice ever since. Some professors do not do that; they ignore the graduate student. . . . In addition, in the last few years I seldom put my name on the paper first, although I have done so once or twice in the last year when the subject of research was something I have been deeply involved in over the last few decades and it's just coincidental that some student happens to come into the picture. But if it's a Ph.D. thesis I always put the student's name first.

In such situations the number of authors of papers and the order of their names will accurately indicate the relative magnitude of the contributions of each author. As groups become larger, however, readers must possess additional knowledge to determine who is actually responsible for the research. A theoretical physicist who had done most of his work in an industrial laboratory indicated that the number of authors of a paper was not a real indicator of collaboration. "Sometimes we just have discussions," he said, "and just as a matter of sociability they put in all the names. But people in the field know who is responsible for jointly authored papers." Such "inside knowledge" is also necessary when scientists order the names of the authors of a paper in an idiosyncratic fashion. For example, an eminent theoretical physicist whose name begins with "A" told about some of his most important work:

> We had many students and many papers. Our first one, the one with five authors, was signed antialphabetically—I had made a rule that I wouldn't have my name first on joint articles. So a grad student's name came first, and in the histories it will be him *et al.*

Since persons may sign articles when they have contributed little or nothing to the research, authorship becomes even less meaningful. Probably scientific administrators in industry are the chief offenders in this regard, although individuals of lower status may also do so.[30] A physics student was asked if publication credit accurately represented the individual's contributions to the work. He replied:

> I don't think so. There is sort of a code of honor; if a person becomes so involved in other things that he cannot participate in experiments, he drops out. This doesn't always happen. Some individuals in our group are [stationed at a different site some miles away] and hardly ever show up here for experiments, but their names appear on papers.

When specialties are small, "people in the field" may know who is responsible for jointly written papers and whose names are included or placed first out of courtesy. Even this would imply that informal modes of recognition would become important at the expense of formal modes of recognition; as suggested in Chapter I, this might lead to the creation of isolated groups that would deviate from the central norms and goals of a discipline.[31] When specialties become very large, as nuclear physics is today, this informal manner of selecting for recognition some of the co-authors of a paper is no longer effective. Then, groups rather than individuals receive recognition. When this happens, the individual is recognized by his participation in the group, and his distinctive contributions may no longer be known by the larger scientific community.

This tendency toward individual anonymity is most marked in experimental physics and was illustrated in 1962, when *Physical Review Letters* published a paper signed by three scientific institutions. The participating physicists were not even mentioned in a footnote. At the end of the letter, valuable help was acknowledged, again without naming individuals but instead other scientific institutions.[32] In the same issue of the journal, another letter was published on the same subject, the discovery of the anti-xi particle; this was signed by seventeen authors from two institutions. In such a large group, the individuals remained about as anonymous as in the first letter. The editor commented on these papers:

> From these and from previous multiple-author papers it becomes clear that in such cases the role of the individual researcher is almost impossible to evaluate. The success of experiments of this kind arises from the combined effort of a large number of workers whose names cannot all be cited in the byline or the acknowledgement. . . . If this trend continues, and this seems to be inevitable, it will have a profound effect upon physics as a profession. In the past, physics research was a rather individualistic occupation with, as principal reward, the recognition by one's colleagues. It attracted those who believed that they could make a personal contribution to its progress. In the future it may require a different type of person, one who gets satisfaction from a cooperative achievement, in which he may eventually arise to a key position.[33]

The practice of multiple authorship weakens the social control exercised by the scientific community through the award of recognition for published contributions. The locus of control is shifted to the formally organized research establishment. Other aspects of the complex organization of basic research have the same consequence.

The Complex Organization of Basic Research

Thus far, this chapter has been concerned primarily with what might be called the traditional organization of basic research in universities. It has assumed that the scientist is essentially free to choose problems and techniques as he pleases, that he is free to accept or refuse collaboration with others, and that the university includes only two research roles, the faculty member and the graduate student.

The traditional forms of co-ordination in science are analogous to medieval forms of economic organization. Free collaboration is similar to the partnership, and the professor–student relation is similar to the master–apprentice relation. These forms of organization of basic research appear to be changing rapidly in many fields. Just as the modern corporation has supplanted free partnership and apprenticeship in industry, so a more complex form of organization may be supplanting free collaboration and the professor–student association in science. Both changes involve the development of a more complex form of organization, the separation of the worker from the tools of production, and greater centralization of authority.

Three factors lead to these results. First, scientific facilities are becoming far more expensive, and access to them and to research grants has become a critical problem to many scientists. Second, more research is done in "interdisciplinary" areas requiring the skills of persons from two or more disciplines. Third, modern scientific techniques and instruments require skills beyond those of a single individual; scientists now require the technical assistance of persons with a doctoral degree or its equivalent. These factors are related—rare and expensive instruments often require technicians with doctoral training, for example—but the effects of each on the conduct of research can be considered separately. The problem of access to research facilities is best divided into the problems facing the scientist who applies to a grant-giving agency for research money and the problems facing a scientist who wishes to gain access to an existing facility.

DEPENDENCE ON GRANT-GIVING AGENCIES

Universities themselves provide only small amounts of money to scientists for research. More than 70 per cent of university research is financed by federal government agencies,[34] and foundations also make large contributions. These agencies do not usually solicit research or attempt to direct it; instead, they approve or refuse requests for grants made by individual scientists employed in universities. Nevertheless, they have a kind of veto power and can be expected to influence research decisions made by scientists. As Richard Colvard has pointed out,[35] most foundations deny the imputation that they do or would seek to control the choice of problems and methods by scientists, and some government agencies might make similar assertions. Foundations, far from undermining the autonomy of the scientific professions, were in danger of having their own autonomy undermined by the professionals. Some scientists, while possibly agreeing with this, have stated that the very form by which grants are given has bad effects on the conduct of research. Most foundations and government agencies use the "project system" whereby the applying scientist describes what he hopes to do and how he hopes to do it. The critics claim that this prematurely commits scientists and inhibits the flexibility of research, and they suggest that grants should be awarded to "the man" rather than "the project." Kidd argues that this is a mistaken view:

> [T]he difference between form and substance, between words and action, is a social and political device serving an essential function. Stress on the substance of research rather than on individuals is perhaps the only way of showing laymen who provide the money how the research expenditure serves a social purpose; and stress on the significance of research proposals to those who apply for money is perhaps the only means of operating the system without generating explosive pressures from within the world of science. Within this somewhat ritualistic performance, the capacity of individuals and groups is in fact given much greater weight than one would assume from the written policy statements of the federal agencies.[36]

That is, the project system manifestly stresses norms emphasizing the importance of research rather than the status of the man performing it.

Scientists interviewed in this study affirmed that their research decisions were not influenced by the prospect of getting research grants. The work of theoretical scientists is seldom influenced by the presence or absence of grants, yet even the laboratory scientists interviewed felt this way. For example, a biochemist said:

> Whenever you apply for grants, you must present some program. The Public Health Service allows you to change it readily. . . . Proposals can be very broad; if you ask for a grant regarding the metallic components of nuclear proteins you can do a great many different things.

These statements are corroborated by surveys of scientists. In Wenkert and Fredrickson's sample of 166 University of California scientists, 82 per cent had their current research supported by grants; only 10 per cent said they usually found it fairly difficult or very difficult to obtain research grants; and 49 per cent said they usually had no difficulty at all in obtaining them.[37] Scientists at less renowned universities probably have more difficulty obtaining grants; however, the leading universities do most of the basic research and the best basic research.

The scientists referred to up to this point were speaking as individuals about relatively small grants for relatively short periods. When more money is involved, the style of scientists is more likely to be affected by financial needs. In Berelson's sample of university faculty members in the sciences, 50 per cent agreed that "too much of what the graduate school does, and how it does it, is adversely affected by the sources of funds, e.g., contract grants supporting dissertation research in the sciences." (Scientists in lower prestige universities, and those who were relatively unproductive, were more likely to agree with the statement.[38]) If large investments in research facilities are being sought, the scientists in charge of institutes must exert great efforts to persuade grant-giving agencies of the desirability of investments. Speyer has argued[39] that the social rationalization of research leads to the confounding of research with politics; more of the time of leading scientists is spent on purely administrative and political tasks, less on imaginative work. Organizational activities lead to deviation from norms of scientific conduct: articles are planned before findings are arrived at, the same findings are reported in several places, and trivial findings are given exaggerated importance.

More recently, the same arguments have been advanced by Alvin M. Weinberg, Director of the Oak Ridge National Laboratory. He finds three reasons why "Big Science" tends to ruin science. First:

> [S]ince Big Science needs great public support it thrives on publicity. . . . If the serious writings about Big Science were carefully separated from the journalistic writings, little harm would be done. But they are not so

> separated. Issues of scientific or technical merit tend to get argued in the popular, not the scientific, press, or in the congressional committee room rather than in the technical-society lecture hall; the spectacular rather than the perceptive becomes the scientific standard.

That is, expensive science tends to orient the scientist away from the community of colleagues to the larger but unqualified community.

Second, scientists come to measure prestige by how much money they spend, rather than by the importance of the results they obtain; "one sees evidence of scientists' spending money instead of thought."

Third, Weinberg argues, large expenditures and large organizations mean that there must be more administrators, and, "Unfortunately, science dominated by administrators is science understood by administrators, and such science quickly becomes attenuated if not meaningless."[40]

Thus, the dependence of leading scientists, those responsible for the efforts of large groups, on grant-giving agencies leads to their becoming politicized. This is not necessarily restricted to the leaders alone, since junior scientists may model their behavior after their seniors. W. H. Whyte claims that extensively organized research "compounds the younger men's interest in the externals of research rather than the content of it."[41] Although the interviews conducted in the course of the present study offer no support for these propositions, they may be true for large segments of the scientific community. Scientists may come to expect that norms of independence are irrelevant, and they may become less responsive to their community of colleagues. Tendencies in the same direction may develop in large organizations built around expensive research facilities.

GAINING ACCESS TO RESEARCH FACILITIES

When research facilities must be shared, decisions about their use are almost necessarily centralized. Particle accelerators, large radiotelescopes, low-temperature laboratories, and similar facilities involve some deviation from the norms of independence in science. My interviews were not concentrated in organizations with such facilities, so the following remarks are somewhat speculative. (Of course, most scientists are not now, and probably will not be, clustered around such instruments. Most experimental physicists are not now doing research with large particle accelerators, for example.) In some places, such as the University of California Radiation Laboratory or the two-hundred-inch telescope at Mount Palomar, the instruments are formally linked with a particular university. In others, such as the Brookhaven Radiation Laboratory, a combination of universities and public agencies operates the device. Usually the larger scientific community is somehow involved in the board of directors of the laboratory, and scientists from many universities are able to gain access to the instruments. I am concerned here not with variations in the formal organization of such laboratories but with the autonomy of professional scientists in one of them.[42]

This organization is built up around a number of high-energy particle accelerators. The facilities are used primarily by the staff of the organization, although physicists from other universities use them about one-seventh of the time. The experimental physicists on the staff are divided into about eight groups, each usually headed by a university professor. The groups are fairly large; the smaller ones may include a professor, a couple of assistant professors, a few postdoctoral fellows, and a half-dozen graduate students, while the largest groups include more than fifty physicists. These numbers exclude technicians attached to the groups, such as film scanners, computer programmers, etc., as well as the technical staff of the organization, such as accelerator crews, who are not attached to any group. The division of labor is complex. A student said:

> We recently made a list of people whose assistance we should acknowledge for the thesis, and it came to around fifty. This included some machinists, scanners, X's group, which loaned us a bubble chamber, someone who helped on IBM programming, and so forth.

Group leaders prefer to have technicians attached to the group, but this is often uneconomical.

The group leader is essentially autonomous, save that he must get the agreement of others to use the facilities. Experiments are suggested and planned within groups, often by junior men. A group leader said:

> I'm pretty much a king, in a constitutional sense, of the group. The people have ideas, it's their work. . . . I spend most of my efforts in trying to keep [the group] working. . . . I see them every day and talk to them. . . . [This is] completely informal, as informal as it could be.

According to an executive in the organization:

> The young men do get involved in planning experiments. They often make the presentations before the research committee, because they often know more about what goes on than do the senior people.

However, he also said, "The character of the group is decided mostly by the group leader," and some groups might be quite authoritarian.

Some students and junior staff chafe under the restrictions of research in complex groups. A solid-state physicist said his specialty recruited many such students. Although physics students are generally oriented to fundamental research, which usually means particle physics, some, he said,

> react strongly to group research and go into solid-state physics. . . . Students are somewhat reluctant to talk about their preferences. They like particle research, but if you can draw it out of them they complain about the group research at the [accelerator organization]. . . . You don't have to be anti-

> social to be driven away from the [accelerator organization]. Work there runs counter to what students have learned in courses. . . . I should correct a possibly false impression: for the staff in the [accelerator organization] there is not much group research. Each man can work on his own. But, for the student, it is group research; it takes a long time before he can do anything on his own.

In this organization, the primary threat to the scientist's independence arises from the shortage of time on the accelerators. A committee of group leaders and theoretical physicists schedules time on the accelerators after reading descriptions of proposed experiments and hearing them defended orally by group leaders or junior staff. The schedule is then approved by the group leaders, the laboratory director having final authority when there is disagreement. While formally each proposed experiment is decided on its own merits, it is likely that each group will be allocated some portion of the available time—sentiments of equity enter into decisions about the allocation of time.

Competition between groups for accelerator time leads to strains. As an executive said, "The leaders are individualistic, and there are many disputes between them." For the most part, groups are differentiated according to the techniques in which they specialize, such as bubble chambers or film, but also according to substantive interests; consequently, competition between groups comes to be seen in terms of which techniques are most valuable or which experimental topics are most interesting. A student saw it from this perspective:

> In this lab a couple of groups are closed and resent it if other groups do experiments in what they conceive to be their area. Violent conflicts may arise if another group tries to do an experiment in their area. Now this kind of conflict is intralaboratory. There are also interlaboratory conflicts. There is much professional jealousy—this is a highly competitive business. . . . One professor . . . tried to just about monopolize use of the X accelerator, using very petty techniques in the process. Finally, some top professors had to tell him off.

Most of the leading physicists apparently avoid overcommitment to a particular technique or experimental topic. A group leader said:

> Disputes like this have emotional roots. Some people think only certain techniques should be used, that all experiments should be done the same way. They are emotional. I personally feel that the only limits to my techniques are what I know and can command. This is not a universal feeling; some men are enamored of a particular technique or machine. . . . Some are fanatics. Others shun them. People are uninterested in hearing things of this sort.

Usually such disputes are handled within the laboratory, "settled behind the scenes," as one informant put it.

Although certain research decisions are made collectively, scientists in this organization have considerable autonomy, either as group leaders or as junior scientists. Despite this, work in the organization could conceivably lead physicists to orient themselves more to the organization and the leaders in it than to the international community of physicists. Group leaders may have to spend much of their time persuading one another, engaging in essentially "political" activities, and they will be able to do this because they have professionally trained assistants to whom they can delegate technical responsibilities. Thus, a rather bitter graduate student answered the question, "Who is the best experimenter around here?" by saying:

> I don't consider faculty members experimenters but, rather, administrators. I don't think, for example, that X put five hours in on the experiment for which he got the Nobel Prize. . . . I don't quite know how to think of the faculty.

The junior men, on the other hand, must be oriented to getting the approval of group leaders. Since groups are highly competitive, loyalty to the group is probably demanded. Group leaders are usually full professors, relatively older men (although many physicists become full professors at young ages), which gives the old critical veto power. In the past, many great physical discoveries have been made by very young men; e.g., Newton did much of his most famous work at the age of twenty-three, and one of my informants did the research for which he received the Nobel Prize at twenty-six. The power of the old over the young may present a problem for physics, although my informants did not think so. The physicist who had won the Nobel Prize while young said:

> Young men seem to take the new situation readily. They don't seem to know that physics could be different. They grew up in it. I think that if all physics should get regimented and all physicists became members of groups, that would be bad—but there are many individual workers now . . . there are enough individual workers, and theoretical physicists still work as individuals. It is not an unhealthy situation.

And he cited the case of a young experimental physicist, recently a Nobel-Prize winner, who had avoided attachment to a large research organization, preferring to work in a university without one and doing his prize-winning work there.

The suggestions that members of this large research organization may become oriented primarily to *organizational* rewards and pressures, rather than those of the community of physicists, is a projection from the real to the possible. In fact, the principal cause of internal strains in this organization has been the orientation of group leaders to the wider community of physicists; many of them are strong candidates for the highest honors awarded physicists, and they know it; for them, access to scientific instruments is one of the most important things in life. The dangers mentioned here will become more severe if and when physicists in the organization become oriented more to organizational harmony than to their status in the scientific community.

INTERDISCIPLINARY RESEARCH

Applied research tends to be interdisciplinary. Medical research requires joint efforts by physicians, bacteriologists, biochemists, analytical chemists, physiologists, and other specialists. Chemical engineering may bring together organic chemists, engineers, physical chemists, electronics specialists, among others. Interdisciplinary research is less common in basic science. However, free collaboration between scientists from different disciplines occurs frequently in some problem areas, and in the recent past a number of research organizations have been established to concentrate the skills of scientists from a variety of disciplines on purely scientific problems.

The free collaboration of scientists from different disciplines is similar to the collaboration of scientists from the same discipline when it involves a division of labor. It begins informally, as a rule, and neither participant has authority over the other. For example, a biochemist in my sample had collaborated with a virologist from another department:

> I knew him. We were friends before, and he had some very interesting problems of biochemical importance. . . . He was asking about the biochemical aspects of this problem. It grew up, and we had a postdoctoral fellow in common. [The virologist] worked on the more biological aspects and I on the more biochemical aspects. . . . There was quite a lot of collaboration in this particular case so far as the actual operations were concerned. The growing and infection of the plants and the harvesting of the viruses was done in their facilities and with their methods, then the biochemical work was all done in our laboratories—we looked at several of the physical characteristics of the viruses, and the most chemical thing we did was physical chemical, finding out by debonding how many groups there were in the virus.

Sometimes one partner is more of a consultant to the other. Statisticians, for example, are often consulted for the statistical design and analysis of studies by other scientists, and this frequently leads to joint authorship of papers. In such cases the authority relations are likely to be ambiguous, and the different backgrounds and values of the participants may produce strains. Statisticians are often harsh critics of the research designs of other scientists, who, to achieve their own goals, may resist taking the advice of statisticians. A statistician who often served as a consultant to medical researchers agreed that playing the critic's role did not endear statisticians to other scientists:

> We must handle people gently in the course of consulting in order to avoid driving them away. We have different approaches to this problem. I think it is important to make constructive criticisms in order to get people to do better work in the future; so I'm willing to avoid mentioning some serious defects in the research for this purpose. Others will launch devastating attacks, to teach the researchers their lesson once and for all. . . . The [agency for which I am a consultant] has a policy of trying to get statisticians in on experimental design. However, the statistician usually has a status far below that of the nonstatistician project director, and for this reason it is hard for him to give suggestions for the research design that will stick.

When interdisciplinary research takes place within complex research organizations, participants are unlikely to have equal formal or informal status, which would produce strains in the relationship, especially since it is not easy to terminate the relationship. The differences in background, status, skills, and values that participants bring to the group could lead to a struggle for power.[43] This is a serious problem for the head of a medical school department in which most of the work is done by interdisciplinary project groups:

> One of my major tasks is getting physicians and Ph.D.'s to work together. . . . [Tension] between clinical and nonclinical work is endemic in American universities, partly because of the unusual status of the physician in our society. . . .[44]

There are many difficulties in getting these groups to operate smoothly. When there are struggles for power, research activities probably suffer. Relations of power become involved in making decisions about the selection of problems and techniques and the necessary tests for the validity of results. Each kind of specialist approaches the problem area from his own perspective and is often incapable of understanding the approaches of others; he may interpret the arguments of others as devices to win power, and they may be precisely that. Perhaps a "least common denominator" approach comes to be used in the selection of problems and concepts. When this happens, significant theoretical findings are unlikely to be obtained.

In the long run, such struggles for power lead either to (1) clear status lines, where one participant becomes a Ph.D. technician (see the following section), or (2) accommodation through the institutionalization of the particular interdisciplinary area.

When clear status lines develop, a scientist from one specialty comes to have almost complete autonomy; he becomes primarily responsible for problem selection and the choice of techniques, and other specialists use their techniques to solve problems set by him. (Of course, groups may be headed by more than a single scientist; the assertion is that scientists representing a single specialty become dominant.) Thus, physiologists may select research problems and the major techniques used in solving them, employing statisticians only to assist in solving certain set problems. Physicians in projects concerned with the physiological effects of radiation may employ physicists in a similarly subordinate role. An experimental physicist who engaged in such "interdisciplinary" research disliked it:

> I was an instructor in radiological physics at X university, but it was not my *forte*. . . . While I was there I was more or less a high-grade technician for M.D.'s. . . . I went into the field not by choice but by force of circumstances. . . . No one wants to be a high-grade technician; I got out of it quickly.

Thus, one form of the institutionalization of interdisciplinary research is the creation of clear lines of authority within research establishments. Another form is the creation of a new specialty. Then, in a sense, the research is no longer interdisciplinary but takes place within a new discipline. This subject will be discussed in greater detail in Chapter IV. Here it is sufficient to note the rapid development of such fields as biochemistry and biophysics. The field of biophysics has developed since World War II. Initially it was interdisciplinary research, bringing physicists and biologists together. Now it has become a discipline in its own right, with specialized journals, scientific societies, and departments in which students are taught the subject.

Interdisciplinary research is often necessary to solve practical problems. As such problems arise, interdisciplinary groups are established, which are dissolved on solution of the problem. Many industrial scientists are successively attached to different interdisciplinary project groups throughout their careers.[45] In basic research, on the other hand, interdisciplinary research seems to be transitory. (This is necessarily hypothetical; we need studies of the development of interdisciplinary research groups that conduct basic science.) It occurs either in free collaboration between professional scientists or in rapidly changing research groups. Until it becomes institutionalized, it is probably an inefficient way of conducting research.

THE PROFESSIONAL TECHNICIAN

It was argued earlier that professional scientists working on the same research problem are inevitably either competitors or collaborators. Authoritative direction can be given in a group that includes graduate students and technicians only because such persons are not defined as professional scientists. "Professional status" has thus implicitly been equated with professional training, and it would follow that there can be no such thing as the professional technician. As a generalization, this is clearly mistaken. In *applied* research many professionals in the sciences place their skills at the disposal of organizational superordinates: others select their problems and methods and define the adequacy of problem solutions.[46] But to what extent is this true in *basic* research?

In a sense, to the technician all research is applied research. He places himself at the disposal of others, allowing them to direct his activities (within a "zone of indifference," to use Barnard's phrase) in return for extrinsic rewards.[47] The technician can become alienated from his work; he will freely sell it for a price. He tends to be oriented toward means and develops an interest in techniques as ends in themselves. To the technician responsible for producing a beam of high-energy electrons, it matters little whether the beam is used for fundamental physical exploration or for treating tumors. Such specialists as applied statisticians and some types of engineers are typically means-oriented and will apply the special logic of their trades without considering their appropriateness.[48]

In most sectors of modern industrial society it is essential that workers, at least some of them, be technicians and thereby capable of alienating themselves from their work.[49] This is apparently coming to be true in science as well. Technicians are *deployable:* they do not have commitments of their own that inhibit them from engaging in research work valuable to others. Research organizations that include technicians are capable of greater flexibility and greater speed in bringing resources to bear on problems. Compare the difficulties of the scientist who attempts to induce free professionals to collaborate with him on the solution of problems, to the ease with which the scientific employer of a scientific technician can obtain his assistance. If advanced training up to and beyond the doctoral level is required to obtain skills in such areas as electron microscopy and statistics, persons with such skills will necessarily be Ph.D.'s. And if research requires such persons to apply their skills at the direction of others, research will require professional technicians.

Although the presence of professional technicians may add to the effectiveness of a research organization, some costs are also involved. The technician is alienated from his work products: he is indifferent to the uses to which they are put. That is, the professional technician cannot be expected to be strongly committed to the norms and goals of science; he is paid not to have commitments that get in the way of others. The technician is not responsive to the opinions of the scientific community; he is paid to be responsive to the needs of his employer.[50] If commitment to the norms and values of science is necessary in research groups, professional scientists must be in control to give the group its orientation.

If, as often seems to be the case, there is no clear distinction between the professional technicians and the professional scientists in a research organization, the technicians may set the tone of the whole. Then the norms and values of science may be subtly undermined. The logic of means will supersede the logic of ends and the research effort will come to be a kind of ritual. It is alleged that already many scientists lean in this direction, and a common complaint is that subtle experimental and analytical techniques are used to solve unimportant problems. (Such complaints should not be taken at face value. It is usually impossible in a scientific article to justify the importance of the problem; editors and others who know the importance of it feel such justifications to be gratuitous. Complaints about trivial and irrelevant work are most likely to be expressed by marginal scientists, those not fully integrated into the scientific community. This will be discussed further in the following chapter.)

The utilization of professional technicians has probably gone furthest in experimental nuclear physics. Informants felt that, although the danger of subversion of scientific values was present, it was not serious. One eminent physicist said that he could not delegate his work to technicians:

> Someone must know what's going on, otherwise the experiment collapses. Choosing the experiment, planning it, keeping in touch with it, is the real tough part of the job. . . . The top man has to be a physicist. . . . The technician will not become a top man. There also has to be a contact with theory, and this is also highly specialized and technical. Thirty to forty years ago I could handle all of the theory and the experiment from start to finish. Occasionally I consulted a theoretician. Now theory is very technical. One must often consult a theoretician. The theoretician himself has a group of students and programmers. Then you plan the experiment. This is essentially an engineering project; it takes many months and thousands of man hours. You need an electronics man, a cryogenics man, and so forth. They tell you what can and cannot be done. If you are lucky, you keep control of it. If not, the experiment will be no good.

As the division of labor in physics becomes complex, the experimenter in charge is likely to become an administrator, necessarily devoting most of his time to co-ordinating the work of subordinates, obtaining resources from others, and similar tasks that keep him away from research problems per se. It takes great men to keep on top of the situation, but experimental physics seems to be attracting them. They are aided by the facts that technicians' roles are often temporary and that their efforts are often well articulated with purely scientific goals.

When the role of the technician is conceived to be temporary, or when technicians are expected to be scientists, some of the dangers discussed above are avoided. As long as the career of the technician is defined as temporary, one which will lead to full professional status, it is similar to the student's role: such a technician, like the student, is expected to understand the goals of the research and be committed to them. (Of course, this involves a cost, since the commitment of technicians to goals limits their deployability. This is less so with graduate students, who are assumed to have less competence and are therefore expected to defer to the authority of their professors. The professional technician usually possesses skills his employer does not.) In experimental physics, the Ph.D. staff members are often involved in teaching students, although they are not faculty members. Graduate students work closely with Ph.D.'s on experiments, and some of them seldom see their professorial advisers. This also means that Ph.D.'s are less likely to become technicians but must keep the wider goals of research in view.

More research needs to be done on university centers and institutes that conduct basic research under conditions unlike those in traditional university environments. We need to know how many nonfaculty persons are involved in research,[51] whether the status of the professional technician is temporary or permanent, and, if permanent, how much autonomy the nonfaculty possess. We need to know the conditions under which the means orientation of the technician comes to set the tone for the entire organization. Until such questions are answered, predictions of radical changes in the organization of science in the U.S. have little empirical basis. My impression, gathered from a sample of scientists primarily engaged in research in traditional university environments, is that technicians seldom set the tone in complex research organizations and that most of them are oriented to professional status in the larger scientific community.

Conclusion: The Organized Disorganization of Science

This chapter has been concerned with the problem of the co-ordination of scientific efforts when such co-ordination requires continuing close work involving two or more individuals. First described were the traditional solutions to the problem, whereby scientists collaborate freely. They work together to solve a scientific problem that interests both of them, together prepare publications reporting the solution, and share equally in the recognition given to the work. This type of joint effort is altogether compatible with the theory of control described in earlier chapters—an exchange of information for recognition.

Traditionally, scientists also obtain assistance from their students. When this occurs, the university scientist is primarily responsible for the information contributed to the scientific community and receives most of the recognition given the work, although he may share it with his students, especially if he competes with his colleagues for students. The dependence of university scientists on graduate students may lead them to avoid risky and long-term projects, and, when teamwork with students occurs, free collaboration with other professionals may become more difficult. These are slight disadvantages to teamwork in academic settings, however. (In industrial and governmental research, risk-taking is also inhibited because it jeopardizes the status of the leaders of more or less permanent research groups.)

Other factors are more important in the tendency of modern science to supplement or supplant traditional forms of co-ordination with more complex forms. Much modern research is characterized by scarce and expensive facilities that must be shared by many scientists. This separates the scientist from the means of production: others determine the uses he will make of research tools.[52] The scientist must be oriented to those in charge of research facilities and must persuade them to give him access to the facilities. His readiness to respond to such officials may reduce his readiness to respond to the community of scientists.

In addition, modern research often involves more specialized skills than any single individual can command. Research is often made more effective if some highly trained individuals become technicians, placing their skills at the disposal of others.

Like other professions, then, modern science is characterized by the splitting of the professional role into the roles of the administrator and the technician.[53] Leaders necessarily become politicized, oriented to obtaining funds and access to facilities and co-ordinating the efforts of others. Technicians become means-oriented, interested in performing their specialized skills for extrinsic rewards and uninterested in the recognition given by the scientific community for the attainment of scientific goals.

If this occurs, the information-recognition exchange theory of scientific organization is no longer applicable. Control is exercised by hierarchical authority within research groups and by political powers outside of them. Scientists become more interested in their particular organizations and in the reactions of politically powerful leaders than in the opinions of the wider scientific community. Consequently, the complex organization of science leads to disorganization—disorganization in terms of the information-recognition exchange theory of organization. The organization that may be supplanting the traditional information-recognition exchange will probably be ineffective for achieving scientific goals.[54] Basic research in complex organizations tends to become ritualistic, reflects low commitment to scientific values, and is displaced easily by practical goals. Attempts are made to accommodate the nonscientific donors of research funds, and the technician mentality sees little difference between pure and applied research in any case. When recognition from the scientific community loses its value, recognition will be sought from other users of research. As the norms of independence for professional scientists become abridged, scientists will come to feel less responsible for achieving scientific goals.

Such a transformation of the organization of science is something of a nightmare now. There is little evidence that the norms and goals of science have as yet been seriously subverted by the complex organization of research. Most basic scientists do not work in the contexts of complex organizations, and those who do are usually oriented more to the interests of the wider scientific community than to their particular organizations. The professional commitments of scientists and the informal organization of colleagues are not easily displaced—a fact well known to leaders of industry who would employ scientists to reach nonscientific objectives.

NOTES

1. This was true for a sample of fifty-seven university scientists interviewed by S. S. West, "The Ideology of Academic Scientists," *IRE* [Institute of Radio Engineers] *Transactions on Engineering Management*, EM-7 (1960), 54–62.
2. See Chapter I, note 73.
3. S. S. West, *op. cit.*; and "The Scientists," *Fortune*, 38 (October 1948), 106–112, 166–176. West stated a number of norms like those I have discussed and asked scientists if, under *any* conditions, conformity with the norm should be abandoned. Since he did not specify alternatives, he found, as was to be expected, that most scientists possessed some values they prized more highly than conformity to certain norms of the scientific community.
4. West, *op. cit.*, p. 62.
5. They may contend that prohibited theories are merely "convenient fictions" (see Chapter VI), or they may disguise a prohibited point of view by clothing it in legitimate garb—German psychologists could disguise psychoanalytic points of view with references to Nietzsche, for example.
6. The Nazi and Communist physicists who attempted to use party dicta for developing physical theory were singularly unsuccessful. See Samuel A. Goudsmit, *Alsos* (New York: Henry Schuman, 1947), pp. 83 f., 114 f., 235 f.; and David Joravsky, *Soviet Marxism and Natural Science 1917–1932* (New York: Columbia University Press, 1961), pp. 286–295.
7. Cf. Willard Waller and Reuben Hill, *The Family: A Dynamic Interpretation* (New York: Holt, 1951), chs. 10 and 11.
8. So it comes about that the men whose names begin with the last letter of the alphabet can never place their names first on a collaborative paper. A story is told of Abraham Wald, who wrote many works with others, and was finally delighted to collaborate with Wolfowitz (on the Wald-Wolfowitz runs test) and so place his name first on the paper.
9. Physicists are excluded from Table 11 because they participate far more frequently in large groups with a complex division of labor and considerable centralization of authority; this will be discussed elsewhere. If physicists are included, the correlation has the same form but a reduced magnitude.
10. *I Am a Mathematician* (Garden City, N.Y.: Doubleday, 1956), p. 200.
11. Compare the findings of Festinger and his colleagues regarding the relation between spatial propinquity and social visiting in a student housing community and the numerous studies showing that propinquity has a very strong effect on the selection of marriage partners. Leon Festinger, Stanley Schachter, and Kurt Back, *Social Pressures in Informal Groups* (New York: Harper, 1950), pp. 153 ff.;

and Marvin R. Koller, "Residential Propinquity of White Mates at Marriage," *American Sociological Review*, 13 (1948), 613–616.

12. Theodore Caplow and Reece McGee, *The Academic Marketplace* (New York: Basic Books, 1958), pp. 68–70.
13. Daniel Lang, *From Hiroshima to the Moon* (New York: Simon and Schuster, 1959), pp. 241 f.
14. It is reasonable to expect distinctive types of social relations to be developed among field scientists as a consequence of their frequent isolation from one another and from research institutions, and one obtains the impression that astronomers, ethnographers, and zoologists who study "natural history" are, for all the differences in their subjects of study, similar to one another and different from other types of scientists in a number of respects. However, I have found it difficult to make this theoretically interesting from a sociological standpoint and have little data with which to support my impressions. Some interesting ideas and observations are reported in Bernard S. Cohn, "An Anthropologist among the Historians: A Field Study," *South Atlantic Quarterly*, 61 (Winter 1962), 13–28.
15. Bernard Berelson, *Graduate Education in the United States* (New York: McGraw-Hill, 1960), pp. 199 f. In Berelson's sample of graduate faculty, 29 per cent of the physical scientists and 27 per cent of the biological scientists had held postdoctoral fellowships. *Loc. cit.*
16. *Ibid.*, p. 176 n. In Diana M. Crane's sample of seventy-one biologists, only 37 per cent reported collecting most of the data they used; 25 per cent reported collecting some of their data; and 28 per cent said that others collected all their data and that they participated only in "design and writing." Diana M. Crane, "The Environment of Discovery" (Doctoral dissertation, Columbia University, 1964), p. 159.
17. Essentially the same results are shown in Derek J. de Solla Price, *Little Science, Big Science* (New York: Columbia University Press, 1963), pp. 87–90. On recent trends in psychology, see Mapheus Smith, "The Trend toward Multiple Authorship in Psychology," *American Psychologist*, 13 (1958), 596–599. Beverly L. Clarke has made a study of papers presented at the annual meetings of the Federation of American Societies for Experimental Biology; the results are consistent with Table 14, but they also show no marked increase in the proportion of multiply written papers from 1946 to 1963. "Multiple Authorship Trends in Scientific Papers," *Science*, 143 (1964), 822–824.
18. The difference between the two disciplines in this respect is also increased by the fact that much chemical research is done in industrial laboratories, whereas mathematics is almost entirely a university discipline.
19. Crane, *op. cit.*, pp. 161 f., also reports a high correlation between size of research group and productivity.
20. Thorstein Veblen, *The Higher Learning in America* (New York: Sagamore Press, 1918), ch. III. For a discussion of current views on this problem, see Berelson, *op. cit.*, pp. 102–109, 143 f., 150–153; Harold Orlans, *The Effects of Federal Programs on Higher Education: A Study of 36 Universities and Colleges* (Washington, D.C.: Brookings Institution, 1962), pp. 68–72.
21. Secondary analysis of Berelson's data.
22. See Chapter I, appendix, on the measurement of productivity for the Berelson sample.

23. *Op. cit.*, p. 148. For example, 97 per cent of the recent recipients of Ph.D.'s in the sciences reported receiving either fellowship support or support from a job as a research or teaching assistant. *Ibid.*, p. 149.
24. *Ibid.*, p. 153.
25. *Louis Agassiz: A Life in Science* (Chicago: University of Chicago Press, 1960), pp. 108–111, 312–323, *et passim.*
26. The best example, because so much has been written about him, is Sigmund Freud. See Ernest Jones, *The Life and Work of Sigmund Freud* (New York: Basic Books, 1955 and 1957), II and III; and Erich Fromm, *Sigmund Freud's Mission* (New York: Harper, 1959). Karl Pearson has also been described as something of an autocrat, and he had many painful breaks with students. See his son's biographical memoir: Egon S. Pearson, "Karl Pearson: Some Aspects of His Life and Work," *Biometrika,* 28 (1936), 193–257; 29 (1938), 161–248. Humphry Davy opposed the election of his most illustrious student, Michael Faraday, to the Royal Society for reasons that are not entirely clear. He may have felt that Faraday did not follow closely enough on the lines he had established or that Faraday failed to give him sufficient credit. Faraday himself self seldom worked closely with anyone else, never even having had a research assistant. In general, he is a rather perplexing figure to the sociologist of science—relatively uneducated, a firm adherent of an extremely fundamentalist sect, and a scientist who matured late. See J. G. Crowther, *British Scientists of the Nineteenth Century* (London: Kegan Paul, 1935), pp. 72–77.
27. Roger G. Krohn, "Science and Social Change" (Doctoral dissertation, University of Minnesota, 1960), p. 55. Diana M. Crane, *op. cit.*, pp. 91–102, reported similar results; she found that the sponsor was most likely to be a role model for his students if the sponsor was accessible, enthusiastic, and permitted students to be independent. For similar observations, see Bernice T. Eiduson, *Scientists: Their Psychological World* (New York: Basic Books, 1962), pp. 167 f.; Howard S. Becker and James Carper, "The Development of Identification with an Occupation," *American Journal of Sociology*, 61 (1956), 289–298.
28. Peter L. Giovacchini, "On Scientific Creativity," *Journal of the American Psychoanalytic Association*, 8 (1960), 407–426.
29. Cf. Florian Znaniecki, *The Social Role of the Man of Knowledge* (New York: Columbia University Press, 1940), p. 41; and William H. Whyte, Jr., *The Organization Man* (Garden City, N.Y.: Doubleday Anchor, 1957), pp. 242 f., 248. Charles Babbage presented an argument of a similar kind in decrying the presence of military men among the leaders of British scientific societies in the early nineteenth century: "The habits both of obedience and command, which are essential in military life, are little fitted for that perfect freedom which should reign in the councils of science. If a military chief commits an oversight or an error, it is necessary, in order to retain the confidence of those he commands, to conceal or mask it as much as possible." Exactly the reverse is true for scientists. *Reflections on the Decline of Science in England and Some of Its Causes* (London: B. Fellowes, 1830), p. 58.
30. Since all members of a team, whatever their contribution, may sign papers produced by the team, we should be skeptical of the value of using changes in the proportion of papers in scientific journals with more than one author as an indicator of changes in teamwork. See Whyte, *op. cit.*, pp. 242 f., where the increase is viewed with alarm; George P. Bush and Lowell H. Hattery, *Teamwork in Research* (Washington, D.C.: American University Press, 1953), p. 173; and Price, *op. cit.*, pp. 87–89. Changes in the proportion of papers having joint

authorship partly reflect changes in practices of assigning authorship and changes in the proportion of all research which is laboratory research.

31. As responsibility becomes diffused, scientists may become irresponsible. A biologist noted that a recent taxonomic article in *Science* had fifteen co-authors and suggested that this would lead biologists to abandon the usual practice of affixing the name of a species' discoverer to its biological names in references to it. "In taxonomic papers, especially, the number of authors should be limited, and editors should demand justification for the inclusion of more than three or four authors' names. . . . The naming of new species has always been highly competitive, and the man who first recognizes a species as unique should be entitled to describe it." James E. McCauley, "Multiple Authorship," *Science,* 141 (1963), 578.
32. *Physical Review Letters,* 8 (1962), 257–260.
33. S. A. Goudsmit, "Editorial," *ibid.,* pp. 229 f.
34. Charles V. Kidd, *American Universities and Federal Research* (Cambridge: Harvard University Press, 1959), p. 51.
35. "Foundations and Professions: the Organizational Defense of Autonomy," *Administrative Science Quarterly,* 6 (1961), 167–184.
36. Kidd, *op. cit.,* p. 108.
37. Secondary analysis of Wenkert-Fredrickson data. See also Orlans, *op. cit.,* p. 94.
38. Secondary analysis of Berelson data. Cf. Berelson, *op. cit.,* p. 152.
39. Edward Speyer, "Scientists in the Bureaucratic Age," *Dissent,* 4 (1957), 402–413.
40. "Impact of Large-Scale Science on the United States," *Science,* 134 (1961), 161–164.
41. Whyte, *op. cit.,* p. 248.
42. Observations of the same organization which confirm those presented here have been reported in Gerald Swatez, "Scientific Norms and Organizational Requirements in a University Laboratory," paper presented at the Fifty-Ninth Annual Meeting of the American Sociological Association, September 1964, Montreal, Canada.
43. Sociologists who have participated in interdisciplinary research projects have made many reports of the tensions arising in them. See, e.g., Gordon W. Blackwell, "Multidisciplinary Team Research," *Social Forces,* 33 (1955), 367–374; and Warren G. Bennis, "The Social Scientist as Research Entrepreneur: a Case Study," *Social Problems,* 3 (1955), 44–49.
44. Contrast the statement of a biochemist in the same medical school, one who engaged only in free collaboration with physicians and seldom in that: "I've been teaching biochemistry in a medical school since 1922, and I've always been in a department of biochemistry in a medical school, but I've never been in such a department which had an M.D. on the staff. We have none here. I have never found any problem on this account. . . . I would discount this as a serious problem. . . . Generally where I have been there has been mutual understanding and respect."
45. See William Kornhauser, *Scientists in Industry: Conflict and Accommodation* (Berkeley: University of California Press, 1962), pp. 50–56.
46. *Ibid.,* pp. 62–73.
47. See Robert K. Merton's discussion of social scientists who become technicians in public bureaucracies in his *Social Theory and Social Structure* (Glencoe, Ill.:

Free Press, 1949), ch. VI. See also C. Wright Mills, *White Collar* (New York: Oxford University Press, 1956), pp. 156–160.

48. For example, applied statisticians, and perhaps statisticians in general, have been accused of using the same criteria for testing light bulbs, where statistical methods have been very successful, and for testing hypotheses in pure science, where the same methods have been less successful. Some statisticians recognize this; see, for example, Jerzy Neyman, "Indeterminism in Science and New Demands on Statisticians," *Journal of the American Statistical Association,* 55 (1960), 625–639.
49. Cf. Chapter I, references cited in note 28.
50. For survey data suggesting that technicians are less interested in being recognized for their contributions and less motivated to achieve scientific goals than professional scientists, see Howard Vollmer, Todd R. LaPorte, William C. Pedersen, and Phyllis A. Langton, *Adaptations of Scientists in Five Organizations* (Menlo Park, Calif.: Stanford Research Institute Technical Report, 1964), ch. III; Donald Pelz and Frank M. Andrews, "Organizational Atmosphere, Motivation, and Research Contribution," *American Behavioral Scientist,* 6 (December 1962), pp. 44 and 46.
51. It has been estimated that there were about 40,000 nonfaculty professionals engaged in research in United States universities in 1958. See Orlans, *op. cit.,* p. 83; Richard J. Petersen, "Scientists and Engineers Employed at Colleges and Universities, 1958," *Monthly Labor Review,* 85 (1962), 37–41.
52. Max Weber mentioned the "separation of the worker from the tools of production" in science; see Hans H. Gerth and C. Wright Mills, trans. and ed., *From Max Weber: Essays in Sociology* (New York: Oxford University Press, 1946), p. 131. The critical factor here is rights to access to the tools. Whether or not a physical chemist owns a mass-spectograph is relatively unimportant; what is important is whether he has sole rights of access to it or must share it with others. On such informal property rights in industrial research see Simon Marcson, *The Scientist in American Industry* (New York: Harper, 1960), p. 87.
53. Cf. C. Wright Mills, *White Collar, op. cit.,* pp. 113–115.
54. See the argument in Chapter I, pp. 54 f.

IV

STRUCTURAL CHANGE: SEGMENTATION

Colleague control is exercised within groups of specialists who share an interest in certain aspects of nature, communicate with one another more than with outsiders, and transmit their goals and skills to succeeding generations. There are many such communities in science, and the relations among them make up much of what can be called the "social structure" of science. This chapter and Chapter V will be concerned with the nature of this structure, typical strains within it, and the way it changes.

Specification of Goals

The formal organization of a scientific discipline is responsible primarily for training recruits and maintaining channels of communication. Its most important units are university departments and scientific societies. Because disciplines in modern science tend to be large and heterogeneous, they cannot serve as informal communities in which recognition is sought for and achieved. Rather, each discipline is divided into smaller communities—specialties—consisting of scientists engaged in research along similar lines. The effective meaning of "similar" here is the ease with which a scientists may engage in work on a new and different problem. This ranges from cases where little new training and information is necessary before the scientist can engage in productive research (problems within the same specialty) to those where years of study are necessary before the scientist can begin to make contributions in the new area. As a rule, the latter extreme characterizes differences between disciplines; consequently, there is little mobility of scientists between disciplines, at least as far as basic research is concerned.[1]

While the amount of mobility possible between two specialties within a discipline depends on the specialties involved, disciplines differ in the "average" difficulty scientists face in changing specialties. Mobility seems to be easiest for experimental scientists in disciplines having powerful and general theories. It is hindered when the specialist must command much relatively unsystematic factual material,[2] and it is also low among theorists working with elaborate and ramified theories. Thus, a graduate student in mathematics said:

> It's easy to switch [specialties] early in one's career. . . . The average mathematician [with the Ph.D.] sticks close to his main subject—it's the easiest way out. This practice is not highly respected by mathematicians. The idea is to be able to switch to another subject every ten years or so.

To the question, "Is there much pressure on mathematics students to specialize?" another gave the following answer:

> There is none up to the time you take your [preliminary examination for the Ph.D.]. But at the thesis stage one must specialize. Sometimes this means a long additional period of study. If you want to write in algebraic topology you would have to study it for two years just to get to the problems. There are two types of problems in math, it seems: some problems are easy, once you learn the machinery for solving them, but this is difficult; other problems require no complicated machinery but are quite complex and difficult in themselves.

On the other hand, each of the four physics students interviewed felt that his dissertation research would not "type" him as a specialist. Each was engaged in research on high-energy elementary particles, but two expected to enter different research areas upon graduation:

> [T]he thesis problem doesn't type you. Graduate work is broad enough to make you capable of doing a broad range of things. *Even things like solid-state physics and spectrography?* Yes. Bell Telephone Laboratories made me an offer regarding work in solid-state physics, but I didn't find the job interesting. *If you did go into such research, how long would it take before you could produce research results?* It would take maybe a year to get into it. Maybe I should amplify to show how research groups operate. In this field all work is done in groups. With a new and difficult problem, industry will put together a group including a broad spectrum of backgrounds—mathematicians, physicists, engineers, chemists, and so forth. So one's past experience doesn't limit one too much.

An assistant professor of organic chemistry had recently begun work on a class of compounds. He expected to get publishable research within one and one-half years from the time he entered the area. When asked how easy it was to move into the area, he said: "This area was not too closely related to the one I was working on. It takes at least one or two, sometimes up to five, years to get to the publications stage."

In any case, the decision to enter a new area of research is something of a gamble, in much the same way as is the initial choice of a career. Time and effort must be spent on working into the area, and this will not yield immediate results—indeed, it will not necessarily yield results even in the long run. A common method of reducing the risk is to do research in two or more areas at the same time, a practice facilitated by the fact that university professors are typically required to teach courses in a variety of subjects. It has an added advantage, since the scientist may get "stale" by concentrating on the same problem for too long.[3] Thus, the chemist just quoted was engaged in research in two different areas; the areas were somewhat related, since he used compounds synthesized in one of them as reagents in the other. A statistician interviewed was engaged on a single research problem; he would rather have had more projects but was restricted to one only by a heavy teaching load: "Other people I know do two or three things at a time and shift gears from one to another when they get stuck as I am. Unfortunately, I haven't had time for this in the past."

A study by Roger Krohn of scientists in the Twin Cities area of Minnesota gives the distribution of number of problems for two samples of scientists. About two-thirds of the university scientists, and an even larger proportion of those working in industry, were involved with more than a single problem at the time they were interviewed.

In the sense in which the term has been used to this point, no two

TABLE 17

NUMBER OF PROBLEMS WORKED ON AT THE SAME TIME—MINNESOTA INDUSTRIAL AND UNIVERSITY SCIENTISTS, 1960

NUMBERS OF PROBLEMS	UNIVERSITY SCIENTISTS	INDUSTRIAL SCIENTISTS
1	33%	21%
2	15	12
2 or 3, plus minor problems	—	14
3 or 4	32	33
5 to 7	20	21
TOTAL	100	101
Number of cases	(87)	(43)*

Source: Adapted from Robert G. Krohn, "Science and Social Change" (Doctoral dissertation, University of Minnesota, 1960), pp. 42, 101.
* Eleven industrial scientists not engaged in research not included.

scientists could share the same professional specialty; no two are engaged in exactly the same sequence of research projects. Of course, problems can be classified, and "specialty" will be used here to refer to a category of problems. Such categories are socially recognized: they are used in the organization of scientific meetings, journals, and teaching, and in advertising for jobs. Scientists identify themselves according to their specialties. None of those to whom I spoke had any difficulty in stating the class of problems on which he was working, and each usually knew who else was working on the same type of problem. (Self-identification may follow the identification of colleagues.)

Many levels of classification exist. For example, chemistry is a subclass of the physical sciences; organic chemistry, a subclass of chemistry; "natural products," a subclass of organic chemistry; and steroid chemistry, a subclass of natural products. Within the specialty of steroid chemistry, a group of chemists may be especially concerned with the synthesis of a particular compound, such as cholesterol. Each subclass may be meaningful socially. Thus, organic chemists often operate as a more or less autonomous division within chemistry departments, sessions at meetings of the American Chemical Society may be devoted to steroids, and steroid chemists may engage in informal communication primarily with those few individuals attempting to synthesize a certain compound. Scientists usually describe themselves as members of a highly specialized group; a mathematician, for example, is more likely to say that he is working in point-set topology than in "geometry," simply because the latter rubric is so broad.[4] Furthermore, competition and teamwork usually occur only between members of the same specialized collectivity.

The kinds of informal communication described in Chapter I are likely to occur only between members of the same specialty. Indeed, it is a common complaint that they are unable to communicate with nonspecialists even within the same discipline. Specialties are often small enough to allow almost all members to be personally acquainted with one another. For example, Menzel mentions a zoologist who was informed of a newly discovered mutant, which proved to be of great significance to his theoretical work, by the discoverer, who:

> wrote to me, as an expert on the cytology of that group (of animals) for technical help, and knowing of my interest. *How did he know of your interest?* It is known in the field as a whole. The cytologists are not such a big group; you know what everyone in the field is doing anyway.[5]

A physical chemist interviewed in the course of the present study said:

> I know everyone in the country involved in this research. Not everyone—everyone who does anything worth a damn. Unless someone has entered the field since last July, started on experiments since the start of the year.

These two scientists were acknowledged experts at universities of international repute; marginally productive scientists at marginal institutions may be less aware of an intimate community of specialists. Even among these scientists, however, many may be in direct communication with a few leaders in their specialty, and these leaders, in turn, may be familiar with most of the work being done.

When specialties are very large or when they are rapidly growing, scientists may feel that most of their colleagues are "strangers." Despite the immense growth in the total number of pure scientists,[6] the number performing research within a particular specialty may have remained small enough for informal communication networks to be effective. That is, the number of distinctive specialties may have grown almost as fast as the number of scientists.[7] If this is so, the informal organization of science may not have changed markedly in the last century and a half.

Even as communication in science occurs largely among members of the same specialty, recognition is usually awarded by colleagues in the specialty: the primary locus of social control in the sciences is the specialty. This means that, once a specialty becomes established, it tends to be self-sustaining. The commitment of each scientist tends to be confirmed by his colleagues, and persons outside the specialty who think it unimportant will be unable to change such commitments. In the long run, however, specialties in a discipline affect each other's growth. Their interdependence is enhanced by the fact that specialists are seldom concentrated in institutions but are, on the contrary, widely dispersed.

CONCENTRATION AND DISPERSION OF SPECIALISTS

Concentration of specialists in a single university department is compatible with some of the goals of departments and individuals and incompatible with others. Thus the distribution of specialists represents a compromise among different interests. The actual distribution of specialists reflects these interests, as well as the labor market and the history of each department.

When members of the same university department choose to work on the same problems, they will become either competitors or collaborators. Intradepartmental competition is far more threatening than competition between men in different organizations, for the former inevitably becomes linked with competition for position and promotion, research facilities, and graduate students. As a result, in some fields no more than one person in each specialty is hired by a department. An organic chemist suggested that this tended to be true in his field:

> The American system is for each faculty member to work independently. This has some effects on getting jobs. If a department is considering you,

and if you're working in the same area as one who is already a member of the department, you're unlikely to be hired. For example, X university might be hesitant about hiring me because of my work with organo-phosphorus compounds. In that case they don't know about my interests in that area, and I'd be willing to change them to something else to get the job. While if I were going elsewhere, I would work in the organo-phosphorus area. Similarly, when I came [here] some of the problems in which I had thought of working began to seem less important to me.

This reason for the dispersion of specialists is most likely to exist in fields where each member of the faculty has facilities of "his own," which he and his graduate students use for research. As noted above, such laboratory sciences are not usually conducive to a collaboration between peers that does not involve a division of labor. In other experimental sciences, facilities may be so complex and expensive that they must be shared by different individuals, and more than one person in a specialty is invariably appointed. The chairman of a physics department said:

In most fields one person is not terribly effective. There is a sort of critical size: we would want two or three persons. For example, in low-temperature physics we have two men and each has a research associate and graduate students. They don't work on the same problems but share facilities and talk to one another about techniques. It is the same for high-energy physics. It would be impossible to maintain an accelerator for only one person. In that area we have two tenure faculty members and want a third, and we have several nontenure faculty members. They don't compete with each other significantly; they keep in touch, and there is no overlap in their work.

As suggested in the preceding chapter, competition is limited by more or less formal agreement when facilities must be shared. Those who use the facilities select problems together, either by discussion and compromise, when they are peers, or more or less by direction, when one is senior to the other.

In theoretical sciences, work done in groups consisting of a faculty member and students limits tendencies to engage in other kinds of collaboration (see Chapter III), which thereby increases the likelihood that specialists will be dispersed. Even when collaboration is readily possible, the desire to avoid intradepartmental competition may lead to the dispersion of specialists. The chairman of a mathematics department asserted that this was not a problem, although, he said:

There have been candidates considered occasionally about whom we would say, "There is not really much point in getting him if we can get somebody else that we think is just as good who has different interests." That has been said at times. On the other hand, there is a substantial amount of overlapping of personal interests, and we think this is quite all right on the whole.

In such fields as mathematics, intradepartmental competition is unlikely to be an important criterion in influencing concentration and dispersion.

The reduction of competition and its consequent strains benefits the department as a whole, as well as individual members. The dispersion of specialists also results because it is felt desirable for departments to be "well balanced." Thus, a chairman of a mathematics department said:

> We feel that it's our obligation as a major state university to be reasonably broad and balanced; it's not proper to have big gaps all over the place. On the other hand, we definitely will build up strength in one or two areas.

Being well-balanced is usually defined in terms of areas of research valued by the larger discipline at the time. Undergraduate instruction plays little role in this, for, with few exceptions, it is felt that any faculty member is competent to teach any undergraduate course. Special teaching competence becomes important only at the graduate level, and there it is linked with the research interests of the larger discipline.

To the extent that considerations of being well-balanced affect the distribution of specialists, one might expect unusual and marginal specialties to be cultivated by large departments only. Small departments would feel obliged to have specialists in areas of major importance and would have no room for others. It does not seem to work out this way, however. In the experimental sciences the faculty is likely to be limited by the research facilities available. Furthermore, small institutions are more likely to be affected by interests leading to the concentration of specialists, namely, that of the department in being distinguished in at least one or two areas and that of scientists in having associates with similar (though not identical) spheres of research.

Scientists usually need and benefit from personal contact with persons of similar interests. When substantive disputes exist in a discipline, or when specialties compete for prestige and facilities, the need for support from colleagues is enhanced. A mathematician said that men in his department were not afraid of competition and did not attempt to keep persons with the same specialty out of the department: "[P]recisely the opposite is true here. Here there is competition to get new positions filled by persons in one's own specialty. . . . [Mathematicians] realize that having good people will benefit themselves; they will learn more." This desire tends to the concentration of specialists in a few departments. The chairman of a mathematics department pointed this out: "Every person here would like to have at least one other person who is at least somewhat similar to him, and for the most part we have this. This is almost an essential for keeping a person satisfied, to have somebody who has interests somewhat like his." This interest is more likely to be effective in such fields as mathematics, where collaboration is easily arranged. Experimental scientists are more apt to attempt to avoid duplication of interests among departmental colleagues.

The interest of departments in having at least one or two fields in which they have a reputation for excellence also leads to the concentration of specialists. A chairman of a mathematics department well known for its work in applied mathematics said:

> One reason for our concentration in applied mathematics was historical. Since the department was small, it was decided to do something and do it well. There was a focus when the department was small. . . . Our general philosophy is that, if we go into a new area, it should be something we can do reasonably well. So, if we do, we will hire a good senior man and some junior men to accompany him. For example, we recently entered abstract analysis, and we added both a senior man and a junior man. The same was true in topology. We add by groups.

The "optimum" distribution of specialists among the departments would reflect a compromise among the different goals that have been described. The actual distribution of specialists is affected by these interests, but it is also strongly influenced by conditions on the labor market and the availability of research facilities. Thus, there is currently a shortage of scientists in almost all fields, but especially in mathematics, and even more so in mathematical statistics. A department chairman said that he had little choice in selecting staff:

> Our situation is such now that we try to expand in order to keep up with our losses. . . . Nowadays in mathematics people are moving all over; it's quite unusual. . . . *When you have an opening do you try to fill it with a man who has a particular research specialty?* No. We used to, but now we don't have the choice. We want to get good people, and we all know it is hard to do so. For example, we really want a statistician, but because they are in such great industrial demand we would have to pay more for one of them than for a first-rate mathematician in another field. So, whenever the possibility arises, we decide against the statistician.

In many of the experimental sciences the availability of facilities is the decisive factor. The chairman of a relatively minor physics department said:

> We can only do the things there is money enough to do. So, if somebody came along with a million dollar endowment for solid-state physics, we would immediately change our course toward that direction. We did something like this just three or four years ago. [We were asked] if we wanted a nuclear physics plant, an accelerator and so on. We, of course, said yes, but we were sort of surprised by this and had no plans for it. Well, now we have a nuclear physics lab, a good accelerator and so on, because we were offered it; the opportunity was there, and we would have been foolish not to take it. . . . Since then we have added three faculty members whose interests are in nuclear physics. And some of the people who were already here have shifted their interests to the particular project.

(It should not be thought that the availability of money *dictates* departmental plans. In this example the department added a high-status research area. If money were offered to assist in investigations in an area of physics thought to be of little interest it might be rejected.)

In general the tendencies leading to the dispersion of specialists appear to be stronger than those leading to their concentration; as a result, there are usually no more than a few scientists in the same specialty in a single department, and there is often only one. To the extent that this is true, we might expect the following consequences:

(1) The dispersion of specialists reduces the frequency of collaboration.

(2) The dispersion of specialists means that scientists tend to look for support and recognition beyond their own university. Thus, dispersion enhances colleague control and weakens institutional control. (This proposition may appear somewhat paradoxical, for, in general, the dispersion of elites weakens colleague control. The difference exists because the alternatives in pure science are not elite control or nonelite control but, rather, control by the cosmopolitan elite of specialists or control by the local group of specialists.)

(3) The dispersion of specialists means that the junior scientist is less likely to be closely associated with senior scientists in his own special field. Dispersion thus reinforces conformity to norms of independence.

Concentration seems most likely to occur when research facilities cannot be distributed. In addition, it may occur to a greater extent in nations in which university education is highly centralized or research is planned, such as England, France, and the Soviet Union. It seems reasonable to suggest the hypothesis that, to the extent that concentration of specialists does occur, collaboration occurs more frequently, scientists are more likely to become oriented to particular research organizations than to the wider scientific community, and authoritative relations between juniors and seniors are more likely to develop.

The Hierarchy of the Sciences

Specialties and disciplines differ in the prestige they are awarded within the scientific community and the larger society; these differences affect the ease of recruitment into fields and the rewards and facilities available in each. Usually there is general agreement about the current prestige rank among the specialties.

It should be noted that few scientists in any discipline would make explicit statements about the prestige of a specialty, or an individual, similar to those made here. In science, as in other equalitarian communities, prestige must be inferred from other behavior. It would be considered improper for scientists to express concern about their individual or collective prestige within science. This norm was well expressed in a letter to the editor of *Science* in 1963; the letter also reveals nonconformity:

> I heard an important speaker at the AIBS [American Institute of Biological Sciences] meetings at Purdue describing ecology as being near the bottom of the totem pole of biology. I was incensed, not so much because it was my first love, ecology, that lay near the bottom, but because there was a totem pole at all in biology. . . . There should be no intellectual "peck-order" in biology, yet one exists.[8]

The same attempt to avoid making prestige differences explicit appears in other groups; prestige-ranking must therefore be inferred from other behavior. One must observe how scientists speak of the relative "importance" or "interest" of specialties, differences in the extent to which formal honors are awarded to scientists in different specialties, and various other forms of deference and snobbery.

CRITERIA FOR AWARDING PRESTIGE

Relations between individual scientists tend to be regulated by the exchange of information and recognition. Recognition is given for information, and the scientist who contributes much information to his colleagues is rewarded by them with high prestige. This process can be generalized to the relations between groups. (For example, the ranking of castes in India is correlated with the types of exchange permitted between them. The castes with highest status may contribute food, ritual services, feces, and so forth to all other groups, or at least all other high-status groups, whereas the castes of lower status are far more restricted in the castes to which they can make contributions.) Information produced in one specialty or discipline may be utilized in another. Sometimes the exchange is symmetrical, each of two specialties producing information utilized in the other. In physics, information obtained by solid-state physicists may be utilized by low-temperature physicists and vice versa. These two specialties could have about the same prestige, if judged on this criterion alone. Usually, however, the relation is asymmetrical: information obtained in one specialty may be important in a second, but information produced in the latter may have few or no consequences for research in the former. When this is so, those in the former specialty may claim and be awarded higher prestige. Their work will be published more readily, they will be more likely to receive honors from the larger discipline, and they will more frequently act as spokesmen for the entire discipline.

This form of awarding prestige is manifested most clearly in physics. Physicists are more oriented toward a single, unified theory than most scientists, and, to the extent that they develop such a theory, specialties within the discipline become less distinct. As has been indicated, physicists apparently find it easier to switch specialties than do scientists in other disciplines. Nevertheless, specialization must occur, and the orientation of physicists to a single theory makes it possible for them to order specialties as more or less "fundamental." Research in the most fundamental areas is unlikely to be affected very much by results obtained in less fundamental areas, whereas research in less fundamental areas often consists in the application of fundamental laws to special classes of phenomena. Physics has a focus, then, at any given time; specialties close to the focus receive highest prestige, specialties distant from the focus receive lower prestige, and specialties extremely distant from the focus are not classified as physics.

Elementary particles are, as their name suggests, of fundamental interest. An eminent theoretical physicist was asked whether the focus of physics on elementary particles could not be explained by the social organization of his field. He replied:

> I think the social organization is very important, but it's not the only reason. The mechanism is a social mechanism, the mechanism of prestige, but at the root of the assignment of prestige there is an intellectual criterion. Now, in the case of all of this type of work, the criterion is that of the fundamental nature of the subject. These laws of elementary particles underlie essentially all of science. All of physics is based on these laws; all of chemistry, all of geology, practically all of astronomy, all of biology, are—in the logical sense—consequences of the laws we are investigating at the level of fundamental particles. That has a tremendous appeal, which accounts for the very high prestige assigned to this field.

Even within the area of the study of elementary particles, sometimes called "nuclear physics," there is some specialization. Of greatest current interest are "strong interactions"—interactions between particles of high energies, which may produce "strange particles"; "weak interactions" appear to be less fundamental. As one practitioner in the latter field said:

> Low-energy work, such as my own, is not going to revolutionize physics. However, two exciting discoveries have been made in low-energy physics in recent years: the violation of parity and the Mössbauer effect.

Those in low-prestige areas are often quick to point out or emphasize the few instances in which work in their particular areas has been of fundamental importance. Of even lower prestige, although still classified as nuclear physics, is the study of reactions between nuclei. (Elementary particles are called "nucleons"; nuclei of all but the lightest elements consist of many nucleons, and their study is more complex but less fundamental than the study of elementary particles.) In this area specialists sometimes complain of inadequate interest in their subject.

Other areas of physics are perceived to be derivative, at least potentially, from the most fundamental areas. This is true, for example, of spectroscopy, optics, low-temperature physics, and solid-state physics. Solid-state physics may be considered to involve

> the many-body problem. This is not high-energy physics but is concerned with deriving properties of matter in the large from laws for matter in the small, which are assumed to be known. . . . It is not considered to be fundamental in the sense of elementary particles, but there are certainly many bright people in it.

Work in solid-state physics seldom affects fundamental concepts in the field, yet there are exceptions. Specialists differ in the extent to which they stress such exceptions. Those interviewed managed to mention them but were careful not to brag about them as indicators of possible future discoveries. For example, a theorist in the area pointed out the work

> by Bardeen and his colleagues on superconductivity, as well as some recent work by Nambu of the University of Chicago. Work of this type has had an enormous back-reaction on physics, even that dealing with fundamental concepts; this rather surprised some of those working with elementary particles. But in other areas solid-state physics is more and more becoming molecular engineering, the application of fundamental concepts without the likelihood of causing modifications in them. Some solid-state physicists are very eager to say it is "pure physics." But, economically, solid-state physics is supported by agencies with applied interests, when it is supported.

Sometimes solid-state physicists are treated as second-class citizens by those working in more fundamental areas, a treatment not appreciated even when these specialists are willing to acknowledge their relatively lower prestige. For example, a solid-state experimental physicist said:

> As far as the professional nuclear physicists are concerned, one meets some feeling that solid-state physics is not fundamental, is almost applied, that maybe we're not—that we're an offshoot of physics, not central. This is generally an irresponsible kind of attitude. From other physicists one gets the recognition that both are legitimate areas of physics—one can do as original work in this area as in particle physics. Sometimes theories developed in solid-state physics are useful in particle physics. Sometimes this is played up a good deal by solid-state physicists. I don't want to exaggerate it, however. I don't think it's necessary; I've always thought that both areas were physics.

Finally, certain areas of physics are designated as "classical":

> [T]here are no longer any changes expected in the theory. Phenomena of a certain kind, within a certain range, are adequately described. Problems are ones of application rather than the formulation of theory.

("Classical" is also used to refer to the kind of theory Newton developed. The usual meaning, however, would include some theories less than twenty or thirty years old.) Physics has been remarkable in its ability to slough off theories that have become classical in this way. At present most of the research in classical physics is done by engineers, and little or no prestige is awarded within physics for this work.

In other disciplines, a similar ordering of specialties is possible. The work of physical chemists is more likely to influence the work of organic chemists than vice versa, and organic chemistry stands in a similar asymmetrical relation with biochemistry. In biology, molecular biology stands at the most general level, while fields like physiology and ethology are far less general. However, these fields lack the kind of unified theory that exists in physics. It is harder, in chemistry and biology, to ascertain which areas of research are most fundamental in the sense of influencing other areas while being uninfluenced by them—largely because it is often impossible to apply allegedly fundamental theories to particular systems.

In the most "primitive" sciences, the taxonomy of the subject matter carries over into the taxonomy of scientific specialists, and the establishment of differential prestige cannot be judged on the basis of the exchange of information for recognition. Who is to say that insects are more fundamental than birds, or that entomologists should be awarded more prestige than ornithologists? In these areas of science, other criteria for assessing prestige becomes more important; as a consequence, agreement on a hierarchy of prestige is also less likely.

In addition to the establishment of prestige differences on the basis of an information-recognition exchange, there is another criterion also intrinsic to the social organization of science, which is generally applicable to scientific disciplines. Specialties differ with regard not only to subject matter but also to methods of investigation: some specialties may be primarily theoretical, others primarily empirical. This cuts across specialization by subject matter. In physics the role of the theorist is clearly differentiated from that of the experimenter. In other disciplines (or groups of disciplines), theoretical roles may not be distinguished from experimental roles, but some specialties may have a larger theoretical component than others. For example, the theoretical component of genetics is much larger than that of other biological specialties, although there are few strictly theoretical geneticists.

Specialties with a large theoretical component of demonstrable utility will usually be awarded greater prestige than specialties with a small theoretical component.[9] There are a number of reasons for this, some having nothing to do with the intrinsic importance of research results. For example, theoretical physicists tend to be selected from among the students who do best in ordinary academic work—work that tends to involve understanding and manipulation of theories more than it involves experiments. The fact that the "best students" tend to become theorists carries over into relations between professional specialists. However, the basic reason for assigning higher prestige to theorists is apparently intrinsic to the relations between theory and experiment. Theory *controls* empirical research, whereas empirical research provides *conditions* for the successful application and manipulation of theory. Thus, although the activities of theorists and experimenters are interdependent and the discoveries of each influence the activities of the other, a qualitative difference in the kind of influence results in differences of prestige. The activities of the theorist influence experimenters to select certain alternatives in planning and reporting research, while the activities of experimenters either exclude certain theoretical alternatives or indicate new tasks for theory. In some respects the relations between theorists and experimenters are comparable to those generally found between leaders and the led.

Other criteria for awarding prestige to scientific specialties are not intrinsic to the information-producing activities of pure science. The practical consequences of research in a specialty may enhance its prestige. However, the importance of the ideology of pure science makes this criterion double-edged. In many disciplines scientists whose research may have practical utility take care to point out that this is not their motive for the work they do. (See, e.g., the preceding quotations from interviews with solid-state physicists.) Those with "pure" interests—the label itself reflects the value judgment—often disparage others. Similarly, although the religious implications of research (e.g., the cosmological aspects of astronomy) or the simple charm of accounting for beautiful or puzzling phenomena (e.g., Newton and the rainbow, or the superfluidity of helium) may give a specialty prestige, especially among educated laymen, it is seldom important today in the scientific community. Again, certain specialties are awarded prestige simply because they are difficult: the solution of problems requires brilliance and specialists have opportunities to demonstrate that they are brilliant. Specialties may even borrow the prestige of their tools, such as electron microscopes or high-speed computers,[10] or their social base—for example, their association with agricultural schools may have reduced the prestige of some biological specialties. Generally, however, such extraneous sources of prestige are relatively unimportant in the sciences. They are most likely to be alleged to be important by scientists who feel their own specialties have been wrongly denied prestige.

Philosophers of science might assign "prestige" (under another name) to a branch of science according to the correspondence of its knowledge with the basic scientific values of parsimony, theoretical generality, logical rigor, and intersubjective testability.[11] Such criteria are important, but to the extent that they are important they can be subsumed under the criteria of relative flow of information between fields and the relative theoretical content of fields. Prestige is accorded men and groups, not bodies of knowledge. Research in some bodies of knowledge of undoubted rigor and generality will have low prestige because developments in them have no consequences of importance for other specialties. These are the "dead" fields, such as classical physics.

As long as science is organized as a community in which information is exchanged for recognition, the prestige of specialties will depend on the contributions they can make to other specialties and on their theoretical content. If science were formally organized and scientists forced to work on some problems rather than others, the practical utility of results and the tools and techniques utilized would probably be more important. This may already be true in industrial research.

The criteria for awarding prestige to specialties need not be consistent among themselves, nor is it always easy to apply them. "Status discrepancies" occur, or there may be widespread disagreement with regard to the relative prestige of specialties. Such disagreement has important consequences.

CORRELATES OF SPECIALTY PRESTIGE

Specialties with high prestige usually find it easy to recruit scientists and succeed in recruiting those with most talent. In specialties with low prestige, on the other hand, the recruitment problem may be the most serious one. Since labor is the most important factor of production in science, this is the most important consequence of differential prestige.

The prestige of a discipline probably influences its ability to recruit capable undergraduates.[12] Within disciplines, the prestige of specialties has the clearest effect on recruitment in physics, the discipline in which prestige differences are clearest. The fundamental problems attract most workers:

> Now there is an overpopulation of people on the fundamental problems such as quantum field theory. It is harder to get jobs if one has an interest in such things. Thus universities where such work goes on can pay lower salaries. Even within universities, people prefer to work on fundamental problems, although they could get better jobs in different areas.

This concentration of effort may be considered unfortunate. An eminent theoretical physicist concluded by saying:

> It is an unfortunate thing, I think, this tendency to group all the effort in a very narrow direction. And we may be losing something in that way. . . . More and more people come in [to the study of elementary particles]. Generally speaking, the ones with the most promise continue to go into the high-prestige field. And succeed in it. . . . From an abstract point of view it might be better if they were more dispersed.

Students may be discouraged from going into the more prestigeful fields. Thus:

> [O]nly the better students are not discouraged from going on to theoretical work. And, within the theoretical field, certain areas are more or less reserved for the best students. For example, only the best students are allowed to work on theoretical high-energy physics.

Even within such a specialty as solid-state physics, physicists tend to concentrate their efforts on the problems felt to be most important. Non-physicists who work in closely related areas have also noted this. A metallurgist whose interests border on solid-state physics said:

> Solid-state physicists are now preoccupied with [nuclear] magnetic resonance, and all of them are working on it. . . . After they exhaust magnetic resonance they will all shift their interest to something like low-temperature research with pure metals. . . .

In fields such as metallurgy, where the prestige order of specialties is not as clear, there is less tendency to focus interests in particular problem areas.

While high-prestige specialties sometimes face the problem of attracting too many graduate students, low-prestige specialties face the reverse problem. One such area is applied mathematics. An applied mathematician in a leading department complained:

> [T]he applied mathematician is looked down upon by the pure mathematician. They don't consider his work real mathematics. . . . We get few students in applied mathematics and usually those who do come in are motivated by the desire to get a lot of money; consequently, they are not the best students. My best students transfer from electrical engineering or physics. Another point: mathematics majors don't know there is such a thing as applied mathematics. If they are good, their teachers steer them to pure mathematics.

If this is viewed as a problem in mathematics, where the assistance of graduate students is not usually important to the research of professors, it is an even more serious problem in the low-prestige specialties of laboratory sciences, where such assistance is often essential.

The influence of prestige on recruitment is reversed when one considers the mobility of professionals (Ph.D.'s) between disciplines. Here the movement, what little there is, tends to be from high-prestige disciplines to disciplines of lower prestige—from "hard" fields to "soft" fields. Meier, after talking informally with more than 1,000 U.S. scientists, made the following generalization about "the hybrid scientist" who receives his Ph.D. in one discipline and practices in another:

> Almost without exception he crosses over from a more precise subject matter to a new area which is less precise in its data-gathering and experimental techniques . . . a feeling of intellectual superiority reinforces one's confidence in himself if he crosses over into a less organized field of behavior, but this is missing if the problem were pursued into a more

> theoretical discipline. In the latter instance the curious scientist will try to establish a cooperative arrangement with some sympathetic specialist rather than educate himself to meet those higher standards of criticism and manipulation of symbols.[13]

The scientist who is established in a discipline carries some of its prestige with him when he transfers to another. This is likely to be gratifying if the transfer is to a discipline with lower prestige, but not otherwise.

For similar reasons, scientists working in areas marginal between two disciplines will be highly aware of prestige differences. One such, a physical chemist, responded to the question whether he identified himself as a physicist or chemist by saying:

> When I'm with physicists I say that I'm a chemist. You know, because of the apparent ranking of field, mathematics at the top, physics, then chemistry, if I identify myself as a chemical physicist it sounds like I'm putting on the dog. But I don't say I'm a physicist when I'm with chemists.

But he, and another man working in a related area, thought of themselves as chemical *physicists* rather than physical *chemists*, and they preferred publishing articles in physics periodicals.

The most important consequences of prestige differences between specialties are on recruitment. This implies that potential recruits are free to choose, which is generally true, although such freedom of choice does depend on organizations' provision of positions for specialists and of facilities for research. To the extent that decisions on such matters, which involve the allocation of capital resources in science, are made by scientists themselves, specialties with greater prestige will stand a greater chance of increasing new positions and new facilities. Even when such decisions are made by nonscientists, considerations of scientific prestige may be important; this will be especially true when such organizations as universities and foundations, which are sensitive to scientific opinion, allocate capital funds. But capital allocation is also influenced by other factors, including, especially, the practical interests of donors. This makes little difference when, as at present in the exact sciences in the U.S., the provision of research facilities is not a major problem for most scientists. It may be more important for European science.

Finally, a man's scientific specialty influences his chances of receiving scientific honors. The well-known honors are usually given to individuals in specialties with high prestige. This may be felt to be somewhat unjust by other specialists. A physicist working in a peripheral area said:

> The Nobel Prizes and so on wind up going to people in these very-high-energy laboratories—because a man in one of them may fortuitously make a dramatic discovery, for which a prize is likely to be granted, and that discovery could not be fortuitously made without the high-energy machine. . . . It isn't possible to become famous for discovering a new anti-particle

> unless you're working in such a place. *I suppose there are exceptions in solid-state physics.* Yes, that's right. Shockley and those men did get Nobel Prizes for work in solid-state physics . . . there are exceptions to everything. But the general trend—why look at the last three Nobel Prizes, they were all for high-energy physics.[14]

When scientists receive well-known honors, not only their own prestige but that of their specialty is enhanced. The award of such honors is one way in which the relative prestige of specialties is made manifest; other manifestations of prestige, such as the nuclear physicists' disdain for solid-state physics, are likely to be private and expressed in relatively subtle forms.

The prestige-ranking of specialties is a collective manifestation of the award of recognition to individuals, and it has an analogous social-control function. An individual who pursues goals thought to be peripheral to the aims of his discipline may receive less recognition, and this will tend to reduce his motivations to deviate. A specialty—a group, however loosely bound—the goals of which deviate from the central goals of the discipline, will typically receive low prestige. This reduces its ability to recruit practitioners, inhibiting its growth. To grow and to receive a greater share of scientific positions, facilities, and honors, the apparently deviant specialty usually must demonstrate that pursuit of its distinctive goals contributes to the achievement of the central goals of the discipline. In some cases this is done, usually by presenting specific discoveries to the larger discipline. In other cases scientists may accept the relative status of their specialty; they may concede that although their work is "valuable" it is less important than the work of scientists in other specialties. In still other cases specialists may reject the imputation that their work is "less valuable" but be unable to convince their colleagues in other specialties of the truth of their position. That is, they may challenge the legitimacy of the prestige-order in their discipline. Such challenges may result from using different criteria to evaluate the excellence and importance of scientific results—in other words, they may involve a conflict of values in science. Since the disagreements usually center around the selection of problems, that is, the short-term goals of scientists, the conflict may be specified further as goal conflict. (Although substantive disputes, those involving the alleged truth or falsity of a theory, are often combined with goal conflicts, it is better to consider them separately—see Chapter VI. Challenges to prestige-rankings also may result from disagreements about the choice of methods, and these too are often combined with goal conflict. A special case concerns the relative prestige of theoretical and experimental roles, which is treated in Chapter V.) Before discussing goal conflict, it will be useful to consider cases of orderly succession of values and goals.

Orderly Succession of Goals

Discoveries in science typically generate new problems for scientists. New theories may require extensive empirical research before they receive "adequate" support; even if they may have potential application for a wide variety of phenomena, only extensive study can realize these potentialities. Conversely, empirical discoveries may require the adjustment or replacement of theories, and anomalous effects observed in one type of phenomenon may stimulate the search for similar effects in others.[15]

Discoveries thus lead many scientists both to re-evaluate the relative importance of problems open to them and to pursue new goals. When many scientists do this, the relative prestige of specialties changes. But specialty prestige is not merely epiphenomenal; as the prestige of a specialty rises, students and professionals may become interested in it not only for its intrinsic importance but also for its importance in the eyes of their colleagues. Sometimes such changes of interest appear to be a kind of collective behavior akin to fashions and fads—at any rate, this is commonly alleged by scientists who oppose the changes. That is, sometimes the substantive causes of the change may be less important than prestige. Be that as it may, the study of alleged scientific fashions will indicate the extent to which purely social factors facilitate the orderly succession of goals.

FASHION IN SCIENCE

In its clearest manifestations, such as women's clothing and household furnishings, fashion is the result of two opposed processes, social differentiation and social emulation.[16] Given a system of prestige-ranking, those able to claim high prestige exhibit it by means of a variety of public displays. As long as those with less prestige accept the system, they emulate the fashion leaders to show that they, too, are in some respects superior. As lower orders emulate those above them, the system as a whole tends toward common forms of display, becoming undifferentiated. This frustrates the wish to display status of those in high-status positions, and they are led to select new forms of display that distinguish them from lower orders. Thus, change is inherent in the system, even if there is neither change in its "environment" nor "technological" change in the system. Since there are usually common norms regarding the limits of display—for example, women must wear some clothing, and in the West there must be some indication of a waistline—the change tends to have a cyclical component.[17] There is constant change, but the change never *goes* anywhere, never results in definite accomplishments providing a basis for future changes.

The ideal type of fashion is not perfectly represented even by women's clothing, in which there is a definite evolution in addition to cyclical changes of fashion, and it is certainly not characteristic of science, where innovations, usually made in one period, are seldom abandoned even when the interests of scientists change. However, in a slightly weaker sense, fashion can be said to be an unstable form of collective behavior resulting from social differentiation and emulation, and in this weaker sense it is common in science as in other spheres of life.

Whereas saying a woman's dress is "fashionable" is usually considered praise, calling a scientific specialty "fashionable" is usually intended as an attack on the motives and good sense of the specialists. For example, in summarizing her thoughts about fashion in cell biology, Fell[18] has written:

> Some techniques become fashionable partly because they are difficult, expensive, or (better still) both, and this gives them a certain snob value; they are, as it were, the mink coats of research. Other techniques become fashionable for exactly the opposite reason—because they are so cheap and simple that anyone can use them, and yet they are *new*, and that in itself confers upon them a certain prestige.

She goes on to consider some motives for following fashions that she regards as improper:

> The field is new, and so prestige is to be gained from working in it; but what is even more important, being new, it is likely to attract money from granting bodies. And this, I think, brings us to a rather pernicious aspect of fashion in research. In general, the waves of interest in something fresh that constantly sweep through our world are vital to its well-being and without them research would indeed be stagnant and dreary. But rushing after new things merely because they are new (or what is more commonly termed "jumping on the band wagon") is another matter; it leads to the abandonment of existing lines of work that ought to be carried much farther, and even to contempt for the realities of nature, as in the disdain for structure that was such a regrettable fashion in cell biology a few years ago.

Those who follow fashions deviate from a norm of science, for they lack originality (by the definition of fashion). They claim recognition from others not because of the intrinsic importance of their work but because of its extrinsic characteristics—the kind of people who do it, the novel techniques or instruments they use, and the financial rewards they receive. Deviation in this form may lead to more serious forms of deviation such as falsifying data or plagiarism, because the deviant has come to be less concerned with solving "important" problems than with obtaining immediate recognition from others.

That those who follow fashions in science are defined as "deviant" by scientists is prima facie evidence against the theory of control presented in this work. I have argued that scientists are oriented to receiving recognition from colleagues, and that this influences their decisions, including decisions with regard to the selection of problems. On the other hand, the fact that scientists may be condemned for following fashions indicates that being strongly influenced by considerations of how others will react is deemed improper, is deviant behavior. That is, if scientists conform to norms as a result of this specific form of social control, they are thereby deviants. This paradox was noted in Chapter I in a more direct form; there, it was pointed out that scientists themselves deny being influenced by considerations of recognition. The resolution of the paradox in this case is the same as it was there: the award of recognition as a sanction operates primarily on the level of individual commitments, and whenever strong commitments to values are expected, the rational calculation of punishments and rewards is regarded as an improper basis for making decisions.[19] The paradoxical deviant nature of following fashion in science can also be resolved, in a strictly equivalent way, by pointing out that it represents deviance from the norms of humility in science and that it represents "defective" gift-giving—the contribution of informational *content* without presence of the appropriate *sentiment*.

The woman who dresses for a more or less formal occasion is not expected to be humble; she is, instead, expected to garb herself in a manner appropriate to the context, and beautiful as well. This does not mean that ladies select clothing only to display status or follow fashions, for usually current fashions are firmly held to be beautiful and appropriate. On the other hand, however, the lady is expected to avoid unfashionable garb because it will reflect on her self and her status—it will indicate either her alienation from the status system or her inability to play her role in it. Overt conformity is demanded as evidence that one accepts the system of prestige and one's place in it, and both conformists and rebels accept this as the meaning of fashion. Of course, it is not always easy to conform, since fashions are always in flux; the claim that something is fashionable is not always accepted. Since it is an important matter, validation of such claims is sought, and such validation can be obtained *only* from others. The simplest and most common form of validation is conformity by others to the same fashion. Deviance by others is resented because it makes one's own claims less secure. On the other hand, if they conform overtly, individuals are given considerable freedom to deviate inwardly; lack of commitment to fashions threatens no one. (Indeed, commitment to *a* fashion represents nonconformity to fashion generally, a resistance to changing one's behavior as the behavior of others changes.) Thus fashion in clothing is characterized by overt conformity, inner permissiveness, and the search for social authorities to validate fashions.

It is precisely this type of conformity that is deviant in science. Motivational conformity is expected of scientists, but they are encouraged to be original in their published research work. In claiming that research is important and ought to be recognized, they should not support their claims by references to others, for to do so conflicts with norms of independence and standards for evaluating research in terms of its allegedly intrinsic merits. Therefore, conformity to fashion is disapproved in science but not in dress. This does not mean that nonconformity to scientific fashions is necessarily approved. While such nonconformity may indicate originality, and therefore conformity to higher level norms, it may also indicate lack of originality, the pursuit of research problems which scientists call "dead," "sterile," "unexciting," or "of applied significance only."

Since the assertion that a research specialty is "fashionable" is often a way of condemning it, such assertions are often made by scientists who feel that changes in the prestige-ranking of specialties hurt their own. A theoretical physicist who was skeptical about a type of theory that greatly interested his colleagues asserted:

> X theory is very much in fashion now. . . . I don't know how it is, but fashions appear in physics. Others have appeared in the past and then, after a few years, they are forgotten. How do these fads start? Very often "big shots" endorse a theory. Two or more of them will do so at a conference. Then an impression is generated by them that something has been achieved. Even at the conference, speakers will begin to defer to this. Then they return and report on the new idea. And then a huge number of papers appears on this. Then it lives for a while until a new fad begins and the old one dies out.

But the statements of scientists about fashions cannot be interpreted only as "ideological" assertions, since they are sometimes made by scientists in fashionable fields. A mathematician engaged in algebraic topology said of his field of study:

> A field may be popular because of . . . more popular mathematicians, big shots—if they are interested in a branch, it grows rapidly. Junior mathematicians want recognition from big shots and consequently work in areas prized by them.

Scientific disciplines differ in the degree to which fashion plays a role in them. These differences appear to be determined by the ease with which scientists can agree on the importance of goals and the typical temporal perspectives of scientists.[20] When the relative importance of goals is easily ascertained by generally accepted criteria, or when the goals are given by nonscientists, there will be little play of fashion. In many of the applied sciences, where the goals arise outside of science and the criteria of success are usually given by nonscientists, scientific fashion is perhaps least important. In the empirical sciences, especially those with a more or less rigorous theoretical framework, the goals arise within science, but in many respects they appear to be "given" in the confrontation of theories by "nature." The opinions of particular scientists are therefore less important—scientists need not look at the opinions of particular others to see how important a problem is—and the play of fashion is restricted. But, in the formal sciences, on the one hand, and the empirical sciences that lack rigorous theories, on the other, it is more difficult to state generally acceptable criteria for determining the relative importance of problems; there is little or no confrontation of theory by nature, which makes some types of problems clearly more important than others. In the absence of such criteria, scientists tend to rely more on the direct social validation of their own judgments, and this allows more play to fashion. For this reason we can expect fashion to be important in mathematics as well as in such disciplines as descriptive biology and sociology. The impression given by my interviews and my experiences in sociology is that this is so.

The typical temporal perspective of scientists is the other factor influencing the relative importance of fashion; this appears to work in the opposite direction from the ease of determining the importance of problems. In fields where problems are resolved and new ones discovered in short periods of time, scientists are prepared to change their own goals and become sensitive to the appearance of new problems of general interest. On the other hand, in fields where problems are resolved less quickly, or where new ones are less quickly discovered, a scientist may commit himself to a problem area early in his career and never change his interests. In the former situation scientists will tend to be more sensitive to the opinions of their colleagues and scientific fashions will be more common.

The difference between mathematics and physics in this respect is remarkable. Some mathematical informants asserted that algebraic topology was a fad in mathematics; this area has its origin in some of the work of Henri Poincaré (1854–1912). Other mathematical problem areas considered to be "relatively new" originated around the turn of the century. In physics, on the other hand, experimental research on pions was considered to be "relatively old. Fermi did the pioneering work in 1950–1953." This seems to be the kind of reference point used by the other physicists who were interviewed. The meaning of time for scientists in different disciplines apparently varies greatly.[21]

The two factors determining the relative role of fashion—temporal perspectives and the criteria for evaluating problems—are inversely related, and this tends to reduce disciplinary differences in the importance of fashion. In physics, for which temporal perspectives are probably shortest, scientists find it easier to agree on the importance of problems. Mathematicians, on the other hand, have longer time perspectives but find it much more difficult to agree on the relative importance of problems. In general, however, fashion appears to be more important in the formal sciences than in the exact empirical sciences.[22]

Such assertions are difficult to substantiate. One must almost be a specialist himself to determine whether an allegation that a problem area is "merely fashionable" is a statement of social fact or an ideological assertion made by a competing scientist.[23] The difficulty is not merely one of measurement. The position advanced here is that the objective criteria scientists have for assigning prestige to specialties almost always interact with the kind of social validation of judgments that gives rise to fashion. (The problem only differs by degree from that in other areas where fashion is important. For example, ladies wear slacks today but did not thirty years ago. The change is not merely one of "fashion" but also reflects changes in the place of women in the labor market and other changes in sex roles.) A rise in the prestige of a specialty, perhaps led by noted scientists, facilitates the allocation of effort to the specialty, which in fact may have become of greater objective importance. And the "objective importance" of a problem area is almost always dependent, in the last analysis, on the presence of a body of scientists who find the results obtained in the area useful in their own work. For example, the objective importance of a systematic theory of genetics apparently depended on the development of evolutionary theory to such a point that the solution of the problem of inheritance was critical. Mendel's work was not as objectively important in 1866, when he published his discoveries, as it was in 1900, when his work was rediscovered simultaneously by De Vries, Tschermak, and Correns.

In any particular instance in which the prestige of a specialty changes, it is probably impossible to disentangle the effects of purely technical factors from the effects of "pure" fashion. Comparing the types of problems selected by scientists in the same discipline in different cultures can provide circumstantial evidence for the importance of fashion. For example, although there is considerable formal communication between mathematicians throughout the world—e.g., Russian journals appear in U. S. libraries and Russian abstracts summarize U. S. papers—there is relatively little informal communication, the kind of communication necessary for the social validation of the importance of problem areas. The existence of this informal barrier may permit different fashions to arise in different cultures, and as a result there are a number of mathematical fields that engage many workers in one culture but few in another. For example, informants suggested that little work in algebraic topology was being done now by Russian mathematicians and that there were other areas in which they were far more active than we are. If the allocation of effort depended on purely objective criteria, one would expect to find similar distributions of effort almost everywhere, especially when formal communication is so complete. Since this is not so, the possibility that fashion plays an important role must be given more credence. In physics, on the other hand, there seems to be less difference between the allocation of effort to problem areas in Russia and the U. S.

Despite such indirect evidence for the operation of fashion, it is necessary to make a detailed study of alleged fashions to determine its meaning and the way it works. It is easiest to begin such studies in the field of mathematics. A number of mathematical informants spontaneously made assertions about these things; for example, an advanced graduate student said, "I suppose you've heard that math runs to fads. Right now the big one is algebraic topology." This allegation was repeated to six other mathematicians who were either algebraic topologists or familiar with topological techniques. All agreed that the field was very popular, and two felt it was at least partly a fad. The field is associated with some very well-known mathematicians, and it has an elaborate theory which, while it is difficult to master, is applicable to a wide range of mathematical problems. Topological techniques may be used, rather than others, because they are new, difficult, and associated with a high-prestige specialty. An algebraist confessed he might have used topological techniques unjustifiably:

> There are competing approaches to the same kind of problem. I have had some of that in my own work. . . . Topological approaches have been encroaching on algebra. [A collaborator and I] once proved something using topological techniques, spectral sequences, that could have been proved with algebraic techniques. The question is whether one should build a complex machine just to prove a few theorems, and we were criticized for doing so. Maybe we wouldn't do it now.

When an algebraic topologist was confronted with such allegations he replied, "Let the scoffers find a simple proof, is what I say to that," and proceeded to give examples of proofs in other areas that could not be obtained without topological techniques. He continued:

> People in other fields are upset because algebraic topology is new. As people become familiar with these methods they will not appear to be complex machinery. Compare analysis: it is in many ways complex, but since people have been long accustomed to the machinery they find it easier to deal with the fine points.
>
> Maybe certain aspects of topology are a fad, but not generally. Of course, all of mathematics is a fad, in a sense; it is what mathematicians are interested in, and this may change. A hundred years from now mathematicians may wonder how we could have been interested in certain kinds of things.

Another specialist in this area summarized his feeling that the area was not a fad by saying, "There is a certain popularity to the field, but a fad flares up and then dies down quickly enough, and I don't think algebraic topology will."

This example illustrates the propositions advanced on the preceding pages: (1) The prestige of a specialty is strongly affected by the uses nonspecialists within the discipline may make of its discoveries. Algebraic topology may have many implications for other branches of mathematics (although my informants could think of no nonmathematical uses for their work). (2) Specialties with high prestige tend to attract new workers. (3) Some scientists are attracted to the high-prestige specialty, or use its results, not for any intrinsic reasons but because of their novelty, popularity, and the fact that the use of the results will impress others not familiar with them. (4) Specialists in other areas may condemn those who engage in the newly popular field as following fashions and assert that the field has been given unjustifiably high prestige. (5) The test of a fashion is the duration of the popularity of a field. A merely fashionable field will lose popularity after a short time—the prestigious reasons for entering the field are of inherently short duration if it is merely fashionable. (The prestige is attached partly because of the novelty of the area and partly because of its unaccustomed difficulty to nonspecialists. As the novelty wears off and scientists become more familiar with the subject, these bases for prestige vanish. A similar process occurs in women's clothing, where the initial prestige of a fashion is associated with its high cost and where this prestige declines as cheap copies become available.)

Even when fashion is relatively unimportant in changing scientific behavior, the study of alleged fashions illustrates how the social validation of the importance of a problem serves to facilitate changes in scientific behavior. Fashion illustrates how informal social processes induce scientists to select new problems. An especially important aspect of this process is the role of the leader—in fashion or elsewhere.

LEADERSHIP

"Leadership" may mean a number of things in science, as it does in other areas of social life. It may mean organizational authority, such as university department chairmen or journal editors possess. Scientists may also be called leaders if they have made great discoveries, whether or not they influence the research conducted by others. Finally, there are the scientific leaders whose brilliance and skill lead others to emulate them. Such men influence others, although they may not seek to do so; I shall call them informal leaders. Sometimes they may be similar to the fashion leaders of "society" in establishing patterns for others to follow.

The active process in informal leadership is not so much the desire of leaders to influence others—they tend to respect the norms of independence—as the desire of followers to benefit from the lead of outstanding men. Followers emulate leaders by changing the direction of their research in conformity with the leaders' examples. Two conditions must be met for this to be common. First, followers must be able to change their research problems easily; the barriers between specialties must be low, as they are, for example, in theoretical high-energy physics. Physicists who were interviewed insisted that "theoretical high-energy physics" was their specialty and that problem areas such as quantum electrodynamics could not be considered specialties. The second condition for the existence of highly influential informal leaders is that such persons be able to demonstrate their superior abilities. When this can be done, followers are able to place confidence in the future decisions of leaders; when it cannot be done the leader may appear to be lucky, or to have had only a single brilliant idea. It seems clear that ability is most easily demonstrated in the formal sciences and the mathematical aspects of the empirical sciences; perhaps this is why scientists in these areas appear to establish their eminence at an earlier age than scientists in the less theoretical empirical sciences.[24] Thus informal leadership can be expected to be most prevalent in the highly theoretical sciences and especially in theoretical physics, where there is a tendency for workers to focus their attention on a narrow range of problems.

An eminent theoretical physicist agreed that this was probably true and suggested it as the reason for the difference between his field and mathematics:

> The number of problems [in mathematics] is much greater. . . . In fundamental physics nature sets the problems. It is a problem of finding laws, not of how they work in particular situations. We know many laws and many problems regarding applications of them. But for the laws themselves there is a limited range of phenomena involved and a very narrow frontier. But there is a tremendous backlog of other problems. Thus, only in *really* fundamental physics are there only a few leaders; in other branches this may not be the case.[25]

The positions of such men as my informant are recognized by others. Their published work is quickly emulated, even by scientists who do not know them personally. They strongly influence those whom they see informally. Another informant, a young theorist who frequently met the one just mentioned, said:

> In theoretical physics there are about five people who I would say are really first class, who stand out above the crowd. . . . They have a prestige, have done extremely fine work, and have made most of the great contributions in theoretical physics. The rest of theoretical physics is done by another 200 or 300 people around the world who essentially take the work of these five people and just work for a lifetime on it and also try to discover theories of their own, possibly. But probably five or ten people have revolutionized the field of high-energy elementary particle physics. . . . There are people like X and Y who stand out a magnitude above the good people

> among physicists. . . . We know these people. They are always willing to listen and shoot your theory down. It perhaps happens much more frequently than our shooting their theory down.

The leaders themselves recognize this and may sometimes find it a problem—they, or their students, are unable to develop the ideas they themselves have put forward. They seldom seek to influence other theorists to follow in their steps. One said that if he published an article a "whole flock of articles" would then appear on the same subject, and that he was one of only a few theorists in this position:

> I don't seek terribly much to influence other theorists. Particularly I don't seek out this kind of work that you spoke of, although it's a useful thing. . . . What I try to do in the way of influencing other theorists is to try to reach agreement with other people, leaders if you like, who work on similar subjects from a different point of view. . . . I like to see people agree on things, and I therefore try very hard to find syntheses. My arguments with other theorists are mainly along that line.

Outstanding leaders do not feel that their followers "compete" with them:

> When you get a result, you follow it as far as you can. When you get stuck, you publish what you have. Then others do the work. They don't "follow up" your results—you usually know their results, having obtained them yourself, and you certainly understand them. After awhile, a few years, people find the places where you got stuck, and then maybe they solve them. . . . I influence others by trying to excite interest, not by telling people what to do. I don't know what happens—I work on physics and the world revolves. I have never thought about the size of my audience.

What is true of theoretical physics in particular is also true, to a lesser extent, of physics in general. The discipline shows a remarkable facility for changing its focus to new goals. Most of the topics that interested physicists a half-century ago are pursued now in other disciplines—chemistry and the various fields of engineering. Even problem areas discovered only thirty years ago are no longer part of physics per se. A Nobel-Prize-winning experimental physicist said:

> The most spectacular thing I've seen in my research career was neutron theory. Now, however, the engineers make a neutron technology and so forth. To them we physicists don't exist any more, we're as remote as Columbus. Sometimes I think they don't know I'm alive.

The ease with which physicists can change the goals of their discipline is linked with the structure of leadership in the discipline. While the ease of determining the really important problems makes it easier to spot leaders, the existence of leaders facilitates the orderly succession of goals.

The orderly succession of goals in a discipline is the sum of individual responses to a situation being changed by discoveries. Changes in the goals of individuals are facilitated by the tendency of scientists both to seek social validation of their goals and to follow the lead of outstanding men. The ranking of specialties by prestige is especially likely to influence scientists whose positions lead them to change goals, such as young scientists and scientists in highly competitive fields. Changes in the prestige-ranking of a specialty lead to changes in the ease with which it recruits new practitioners.

The succession of goals in science is not always orderly. Alternative goals may be thought to be incompatible. Scientists may feel that methods and standards for evaluating research in different specialties are so vastly different that they do not belong in the same discipline. Assertions that "this isn't *really* mathematics"—or chemistry, or physics, or biology—are common. In such cases the succession of goals is not achieved in an orderly manner but only as the result of conflict. The earliest manifestation of such conflict is the emergence of "deviant" specialties, groups whose members feel they are not awarded as much prestige within the discipline as their efforts deserve. Such deviation may evoke formal types of social control, and the conflict that may result focuses on such forms of control.

Goal Conflict

There are two kinds of deviant specialties. One comprises those specialties with members who accept the goals of their discipline but believe their specialty is much more important relative to these goals than others give it credit for. The other comprises specialties with members who in effect reject the central goals of the larger discipline and, therefore, the legitimacy of the prestige system in it. For convenience, let us call the former specialists "reformers" and the latter "rebels." Although scientists often say that a specialty is not really part of their science, they would certainly not use the word "deviant" to describe it. The norms of independence prohibit the use of such moral terms in describing relations between specialties; for the same reason, even "specialty" itself would be used by scientists in a strictly technical sense, not to represent a group. "Deviance," "reform specialties," and "rebellious specialties" are all inferences on my part from the observed behavior of scientists, inferences they themselves are unlikely to make.

REFORM SPECIALTIES

The prestige of a specialty is normally determined by the importance of its research for other scientists. This "importance" is inherently ambiguous. Usually past and present contributions, plus estimates of the importance of problems remaining to be solved, are used to evaluate the possible *future* importance of the discipline, and these evaluations are crucial in the prestige awarded it. Estimates of potential contributions, however, are necessarily speculative; members of a specialty may not agree with outsiders about its potentialities. Members may feel that a long view is necessary and that contributions made in the recent past are less important than future potentialities, while outsiders may resist claims to prestige that are not based on "solid" achievements. Furthermore, it is not always easy to evaluate the actual importance of past contributions. If there is disagreement about past contributions or potential contributions, specialists may come to feel that the prestige awarded them is too low. This can be illustrated by considering the problem area of relativity in theoretical physics and the specialty of logic in mathematics.

Theoretical physicists interested in the general theory of relativity, or a unified field theory, often refer to it as an "unfashionable" specialty. Here, as before, "fashionable" has a moral connotation: the implication is that most theorists neglect the field for reasons having little to do with its intrinsic merits. Unlike theorists in solid-state physics or physicists in such areas as acoustics and spectroscopy, who accept the relatively low status of their specialties, relativity theorists may feel that the lack of attention given to their specialty is unjustified; they may feel that various forms of coercion exist to discourage men from working in it. However, even physicists strongly interested in relativity agree that it is unlikely to produce results useful in other branches of physics in the near future:

> It is a very difficult field. One could spend years without getting results. You can't write lots of short papers. . . . Some young people are interested in relativity; they may write theses in it because they are not aware of the state of affairs.

Physicists nevertheless assert that cultivation of this branch of physics has great potential importance for physics. Einstein himself, who was relatively isolated from his colleagues in his later years because of his interest in this specialty instead of quantum theory, wrote:

> I believe . . . that [the contemporary quantum] theory offers no useful point of departure for future development. This is the point at which my expectation departs most widely from that of contemporary physicists. . . . No ever so inclusive collection of empirical facts can ever lead to the setting up of such complicated equations [as are necessary for a theory of gravitation]. . . . Equations of such complexity as are the equations of the gravitational field can be found only through the discovery of a logically simple mathematical condition which determines the equations completely or (at least) almost completely. Once one has those sufficiently strong formal conditions, one requires only little knowledge of facts for the setting up of a theory. . . .[26]

Their rejection of the prestige system of physics leads to the isolation of relativity theorists—they are isolated by others and they protect themselves behind barriers of isolation. Relativity is not a case of a seriously deviant specialty: the specialists apparently do not form a cohesive group.

The definition of relativity as an unfashionable specialty was denied by other theoretical physicists:

> General relativity is not unfashionable. That's an excuse. It is just unproductive. If someone got significant results with it, people would immediately begin working on it. As it is, papers report results of the form "I have shown that Professor X's equations can have only two solutions"—with no physical meaning.

The same informant suggested that Einstein tackled the problem because, "He apparently desired to repeat his revolutionary discoveries and tackled the biggest problem he could find, that of unified field theory."

While relativity theorists reject the prestige-ranking assigned them within physics, this has not produced serious strains in the discipline. In the case of logic, on the other hand, members of the specialty are organized: they form a deviant specialty within mathematics.

Mathematics derives its prestige largely from its being the servant of the sciences; it contributes to them all, but they make no real contributions to it in return—they only provide "problems," which, although perhaps important historically, are hardly contributions in a scientific sense. In addition, mathematics is a highly "theoretical" science; it is a "pure" science; and it is difficult, requiring the highest intelligence. These characteristics, which give mathematics its prestige relative to other disciplines, should also give mathematical logic high prestige relative to other mathematical specialties. Logic can make contributions to mathematics, but the information given in the other direction tends to consist in problems and "techniques," not valid results. Logic is even more "theoretical" than mathematics in general—more abstract, more general, and more rigorous. But this has not resulted in high prestige for logic and logicians within the field of mathematics; mathematicians assert that logic is not important to them, that logicians are not "creative" or "good mathematicians," and that maybe logic is not really mathematics but, instead, "philosophy."

The reason mathematicians give for the low importance attributed to logic—in other words for assigning it low prestige—is that its contributions to the rest of mathematics are few. "Deep theorems in logic only rarely play a part in mathematics. They're a different discipline. We don't talk about the same set of objects. . . ." A graduate student put it more strongly:

> I don't like logicians. Only two or three good theorems have been proved in logic in twenty years, and Goedel proved them. Mathematical results

> should be things that are important in the sense of leading to new developments rather than merely proving what you already know.

More is involved in the relatively low prestige of logic than the alleged uselessness of its results for mathematicians. Mathematicians may resist logic because the logician acts as a critic of mathematical work—as one of them said, "A mathematician looks at a logician as a novelist looks at a grammarian." The logician may question various procedures used by mathematicians. (In the same way mathematical statisticians may criticize applications of statistics and meet resistance from the users of statistics for this reason, and mathematicians may criticize the work of theoretical physicists and receive their denigration for paying too much attention to "trivial" details.) One logician pointed out that Frege, a founder of modern logic more than half a century ago,

> said it was scandalous that mathematicians talked about numbers as their main objects, and no mathematician was able to give a clear definition of what number is. He wrote one or two satirical things, very ironic; he made fun of the various definitions given of number.

When the logician asserts that the mathematician's proofs are suspect because of his use of unproved propositions, the mathematician may respond that "mathematicians go ahead on faith" and that they are justified by their works. The mathematician may go on to assert that mathematical creativity is injured by training in formal logic:

> Mathematicians like to think about their field in terms of content, and they can claim that logicians destroy the content by formalizing it and making it into something mechanical, whereby it loses all its interest and all its elegance. . . . Poincaré is the classical example; he's now past history, but there is still something of the same reaction.

Similar arguments are made in other disciplines by scientists opposed to what they regard as the unnecessary or undesirable use of mathematics and quantitative methods by their colleagues. Usually this fails to affect the high prestige given the most mathematical specialties; in mathematics itself, however, the same arguments are used effectively against logic. Partly, this is because logic is also in philosophy, hence many logicians are not skilled mathematicians. A logician holding a position in a philosophy department said:

> Mathematicians have higher standards than philosophers; [the book by X, a leading philosopher-logician], his major contribution, would make a good five-page article in a mathematical journal. This is one reason why I want to identify myself with mathematicians—they have higher standards.

Logic is also like philosophy in that it has some basic unresolved questions—major points on which logicians cannot agree; in this way it differs from the rest of mathematics.

The logicians who were interviewed felt that mathematicians often disparage logic because they are misinformed about it. One logician said there were two reasons for the hostility of mathematicians:

> At first, because mathematicians regard logic as philosophy—because it doesn't look like mathematics and is practiced by nonmathematicians. Then, when it becomes obvious that it is mathematics—it looks like it, appears in mathematical journals, and so forth, they are hostile because of the threat of innovation. . . . Mathematicians are suspicious if something shows signs of being important that they didn't learn about when young; they feel threatened by such things.

This is associated with the long-time perspectives of mathematicians; they seldom find it necessary to change their special fields of work because all the interesting results have been obtained, and they usually find it difficult to make such changes.

Logicians argue that in the last half-century their contributions have profoundly transformed mathematics and that such contributions will continue to be important:

> Today a great deal of mathematics and the whole of analysis is based on set theory, and set theory is nothing but a developed mathematical logic; today mathematicians proceed in a way quite different from the way they proceeded even fifty years ago. They are now on a sounder basis, but they say, "Set theory is really mathematics." Well, you can call it mathematics or call it logic. It was developed by a mathematician . . . who worked quite alone and was unappreciated by other mathematicians, Georg Cantor of Germany. . . .[27] So, the mathematicians call set theory mathematics and are fairly comfortable with it, but if it were called logic they would feel it was entering mathematics from outside.

As these statements indicate, logicians often insist that logic is important to mathematics and that they share the goals of other mathematicians. The logicians whose work is most purely "logical" tend to identify themselves as mathematicians, whether they serve in mathematics or philosophy departments. The philosopher-logicians interviewed for this study were more interested in the history of logic or in applications of logic to other (nonmathematical) fields; they were less likely to be actively engaged in research in the field of logic per se. Logicians usually said the leading contributions appear in mathematical periodicals, certainly not in philosophical periodicals or even in the *Journal of Symbolic Logic*, the "interdisciplinary" journal devoted to the subject. Thus, insofar as logic is a deviant specialty in mathematics, it is a reform specialty; logicians do not insist on the legitimacy of their nonmathematical goals. Nevertheless, logicians are organized as a group in the Association for Symbolic Logic, they have a journal of their own, and in some universities they form organized groups of the faculty; they possess the ability to act collectively to further their own view of the goals of mathematics.

Relativity theorists and mathematical logicians are alike in rejecting the status awarded them in their respective disciplines and unsuccessfully claiming higher status. In other respects they are quite different. Relativity theorists do not form an organized group and have no specialized communications media. Logicians, on the other hand, have a "consciousness of kind," a society, journals, and to some extent distinctive ways of viewing things that are not required technically by their specialty. Other reform specialties are more or less similar to one or the other of these. For example, applied mathematicians are organized as a distinct group with societies, journals, and some centers of research in universities. Applied mathematics is characterized not so much by claiming high prestige among mathematical specialties as by claiming status as *a* mathematical specialty. Applied mathematicians must combat assertions such as "much of applied 'mathematics' is not mathematics," or not "real mathematics," which are often made by pure mathematicians. The reform they seek is to remove the stigma attached to applied mathematics of being "uninteresting," "routine," or not "real" mathematics. If this cannot be accomplished, applied mathematicians may conceivably reject their commitments to the goals of mathematics, thereby becoming rebels instead of reformists.

REBELLIOUS SPECIALTIES

When the charge is made that the goals of mathematical logic or applied mathematics are not the same as the goals of mathematics, specialists in these areas deny it and attempt to demonstrate its falsity by making contributions that will be regarded as mathematics by other mathematicians. On the other hand, mathematical statisticians, who now form a separate discipline but were once part of mathematics, are more likely to agree with such a charge or to make it themselves. For example, a statistician who had received his doctorate in a department of mathematics asserted that statistics should be an independent department because "statistics is not mathematics—it uses inductive reasoning, decision making, model building, and so forth—we *use* mathematics but are interested in results." This does not mean that statisticians are of a single mind on this subject; those serving in departments of mathematics often claim that their field is part of mathematics. Even those who insist on the independence of statistics state that it is a mathematical science and are proud of the mathematical accomplishments of statisticians. Thus Jerzy Neyman wrote, in a volume commemorating Harold Hotelling:

> The development and dissemination of the excellent theoretical results by Hotelling himself, and of the many talented authors collected and inspired by him, contributed considerably to the establishment of statistics as a respectable mathematical discipline.[28]

Statistics is *a* mathematical discipline and need not be bound by the limited goals and standards of pure mathematics.

Statistics is now emerging as an independent discipline; it is no longer a rebellious specialty within mathematics. A rebellious specialty is formally included within a larger discipline but its members reject the claim that its goals are or should be the goals of the larger discipline. This means that such specialists reject the prestige hierarchy of the disciplines of which they are formally members; the evaluation by their disciplinary colleagues of their specialty is not regarded as legitimate. This usually leads to claims for facilities and rewards that others in the discipline regard as incongruent with the "importance" of the specialty. This, in turn, leads to *organizational* as opposed to *informal* conflict.

The phenomenon of rebellious specialties is common in modern science, especially in modern science in the U. S. For example, many scientists who are formally members of chemistry departments conduct their research on the borderlines between chemistry and physics and tend to reject the idea that their goals are the goals of chemists in general. They resist identification as chemists, for example, by refusing to join the American Chemical Society; they publish in journals of physics, have formed societies and published journals of their own, and assert the need for specialized departments or centers of "chemical physics" in universities. The current developments in the fields of biology are perhaps most interesting in this regard. The classical disciplinary boundaries of zoology, botany, physiology, bacteriology, and biochemistry have become less meaningful as scientists in all these areas have begun to study life processes at the molecular level. These scientists tend to reject the goals traditionally accepted in their disciplines, which has led to recurrent goal conflict between classically oriented scientists and molecular biologists.

In molecular biology the successes of a rebellious specialty have placed their opponents on the defensive, and assertions of the importance of traditional goals have sometimes appeared in print. For example, Barry Commoner, in his address as retiring Vice-President of the American Association for the Advancement of Science, in 1960, noted the "alienation" of molecular biology from the more central biological fields. This is seriously manifested in the choice of research problems by young men:

> Twenty-five years ago, bright young people eager to conquer the world of science were proud to become biologists, to study *Drosophila* genetics, plant taxonomy, or embryology. Nowadays a student with a budding interest in genetics often ends up mating strands of DNA rather than fruit flies, and greenhouses are built to grow plants for the purpose of producing viruses. Bright young biologists, if they are good enough, become biochemists and biophysicists.[29]

The overemphasis of such scientists upon physics and chemistry means that they are unable to contribute to purely biological goals.

> [A]s soon as an interesting and important biological problem becomes susceptible to chemical or physical attack, a process of alienation begins, and the question becomes, in the end, lost to biology. But in each case, the purely chemical—or physical—studies run their course and come to the blank wall that still surrounds the intimate events which occur within the *living* cell. The obvious need is to return home to biology. But now the errant science has long forgotten its home, and the mother is too bewildered by its fast-talking offspring to be very happy about welcoming it back into the family.[30]

This is alleged to be at least partly the result of mistaken factual ideas. Commoner attacks the views of Isaac Asimov, a molecular biologist, who comes to the conclusions that

> life *is* a function . . . of a molecule and not of the cell, that life is *not* uniquely different from non-life, and that the difference between life and non-life is disappearing.[31]

Commoner asserts that if this is factually true, or "even remotely correct, biology is not only under attack; it has been annihilated."[32] He attempts to refute this allegedly *factual* argument in his endeavor to show that the *goals* of traditional biology should not be abandoned. People in general use factual assertions to support arguments for and against particular policies, but the tendency to do so is probably greater in science, where recourse to "traditional goals," "authority," and legal codes is considered illegitimate. As a result, arguments about policies and goals shade into arguments about substantive truth, and in many cases one can make only analytical distinctions between them.

Public expressions of goal conflict like those of Commoner and Asimov, occur relatively infrequently—far less frequently than goal conflict itself—and they can probably be legitimized only by presenting them as factual arguments. This is so because scientists are expected to be independent and respect the independence of others. In fact, most scientists are only too willing to let others go "their own way"—i.e., select their own colleagues, and their own research problems—if they themselves are given the same freedom. Scientists may privately scorn the goals selected by some of their disciplinary colleagues, but they are less likely to express such sentiments publicly or to a stranger. However, the scope of such tolerance is limited; in some contexts the selection of goals by individual scientists necessarily becomes a matter of concern to collectively organized scientists. The most important organizations are university departments, but scientific societies, the journals they establish, and grant-giving agencies may also become involved. When informal consensus breaks down, conflict is "kicked upstairs" and formal organizations become involved.

Organizational Consequences of Dissensus

If scientists agree on goals and on the degree of prestige due specialties, organizational decisions regarding relations between specialties are easily made: a priority schedule can be established for the allocation of facilities and rewards, and this schedule will be more or less accepted by members of the organization. In the absence of goal consensus or a legitimate prestige hierarchy, organizational decision-making is more difficult and less harmonious.

UNIVERSITY DEPARTMENTS

Departments in leading universities are usually organized on a more or less collegial basis, in fact if not in form.[33] The critical decisions in all collegial bodies involve the selection of new members, and this appears to be the principal occasion for conflict among competing specialties in university departments. A legitimate prestige hierarchy may regulate relations among specialties without specified organizational policies. In the absence of consensus, decisions must be made by an authority, majority vote, or some other formal procedure.

Ideally, the selection of new members of a department is simply based on excellence; the personality of the candidate, his social origins, and his specialty are supposed to be of secondary importance. The following statements by chairmen of mathematics departments are typical:

> We add by groups. But we don't have a conscious policy of adding so many people in this area, so many in that. Our emphasis is on getting people of high quality, irrespective of specialty. With older persons specialty counts a little more. Right now, due to losses, we would like to get more classical mathematicians. It is hard. We have no quota: we would like to maintain our reputation, get it better, and make the atmosphere pleasant—and there is nothing like quality for that.
>
> We have certain areas we know we would like to strengthen. On the other hand, when you come to look for people and you find some very good people, there is a strong temptation to take them whether or not they fit into that one branch.

It is often difficult to evaluate the competence of a prospective recruit, for specialists may find it difficult to evaluate work in other specialties within their discipline. Deviant specialties make such decisions even more difficult. Departmental colleagues may agree that a man is a highly competent applied mathematician or logician, but this does not imply, for all of them, that he is a good *mathematician.* Scientists in deviant specialties find it more difficult to get jobs. For example, a theoretical physicist said:

> In most departments quantum mechanics is in control and it is difficult to be interested in relativity. People who tend to work on what *they* think are really fundamental things are less likely to get jobs. And maybe correctly. People who unjustly criticize the lack of rigor in physics may lack competence. But there are some competent relativists. . . . Thus, a thesis in quantum field theory, meson theory, or something like that is definitely the thing to have.

The critical decisions occur when a scientist becomes eligible for full membership in the colleague group, i.e., when he is considered for tenure. At such times the specialty of the individual under consideration may be a dominant consideration. One case was described by an interested observer:

> The mathematics department [at Z university] is conservative. . . . I told the chairman that I understood his department was hostile to logic. He said, "No, we aren't hostile, like we are with algebra, we are just indifferent." They have no algebraic topologists either; they are generally hostile to modern algebra. . . . They have a young Assistant Professor X. He does research in logic but teaches something else. He came up for tenure recently, but they didn't judge him on his merit, they weren't able to, but, rather, on whether they wanted to develop the area. A promotion for him would have been felt to be commitment to take on more logicians later on. They had five assistant professors. They couldn't have promoted them all, then they wouldn't have room for any newcomers; this would have been unwise administratively. So they postponed the decision on all five. With two of them this made sense, since they were quite young. . . . Neither of [the other three] really fitted in with the departmental aspirations.

The chairman of the department in question was a good deal more discreet on this topic. He was asked, "Suppose a member of the department is taken on for work in one specialty and later tries to change to a different specialty. Will any pressure be put on him to make him desist?" He replied:

> The member of the faculty is king, he does as he wishes as far as his mathematical interests go. However, if his interests shift to music, not mathematics, they will be discouraged. There are some borderline areas mathematicians are tempted to go into, and this can cause complications.

From other interviews it seems reasonable to infer that this man would classify logic and certain types of "modern" mathematics in the same line as "music." This particular department is unusual, having attempted to cultivate a style of mathematics more or less abandoned elsewhere. But the problems of those in deviant specialties are fairly general.

With regard to initial appointments, the pressures on those in deviant specialties are impersonal: they simply face a labor market that has a low demand for their skills. On the other hand, the deviant specialist on the scene who is being considered for tenure may be exposed to direct pressures. Given the norms of independence, these must be expressed subtly, but they are expressed. A mathematical statistician employed in a department of applied statistics said he was leaving his position because of the pressures on him to do research in applied, instead of mathematical, statistics. These pressures were exerted by the department chairman:

> [This department] trains people for well-defined professional positions and there are pressures to do research along these specific lines. These pressures are not explicit, but if you aren't dead you feel them.

In the same way, those without tenure can be influenced to remain in specialties by implicit pressures: the young man must publish, and it usually takes considerable time in a new area before publications can be prepared. A physics department chairman said that he had no influence over the research decisions of tenured staff:

> It could be embarrassing if one went into something dull, uninteresting, or even gave up research entirely. With the junior faculty it is different. They are usually here for a short time, three to four years. They must stick to a topic to get anything at all done. . . . If they want to switch afterward, we permit it, but it is not usually in their own best interests. Thus pressure is not applied.

Of course, specialists seldom choose to turn away from a high-prestige specialty to one of lower prestige. Changes in the other direction are more likely to receive greater tolerance. A mathematics department chairman said:

> We had a man who had been Einstein's assistant at the Institute of Advanced Studies and got his degree from Columbia. . . . Quite a number of us sort of supposed that he would interest himself in the mathematical theory of relativity and related questions. . . . He let us know quite soon after he came here that he was interested in number theory. But we didn't really choose him because of his specialty. A few people were disappointed because he didn't want to teach certain courses that we thought he would like to teach. . . . I don't think it would be regarded as proper at all to try to pressure a person to work in a research area that he had decided he wanted to turn away from. At [X University] a professor of mathematics became rather well known for writing a history of the civil war. . . . I took the point of view that the department was not being well served if he gave up mathematics. But such things happenly rarely, and I don't believe in trying to work out a policy ahead of time on them.

Once a university scientist has tenure, it is more difficult to apply formal pressures to induce him to select some research problems rather than others. However, informal pressures can be used. An applied mathematician spoke of the difficulty of getting other applied mathematicians appointed in his department:

> We want an applied mathematician. I'll suggest a man, but I must be careful to suggest one the pure mathematicians consider to be a mathematician. For example, . . . one man in our department turns out much good mathematics —good from an engineer's standpoint; but he is ostracized by the pure mathematicians, considered inferior by them. . . . It's hard to say. Mathematicians don't feel any creativity is involved in his work. He applies standard techniques to new problems. He may be ingenious, but it is not mathematical creativity.

The individual in question mentioned his isolation from his departmental colleagues. He was quite tolerant and good natured, but made some trenchant criticisms about the "trivial" work often done by "pure" mathematicians. However, he felt, with some regret, that his own work was pure mathematics, not suited for application by physicists or engineers.

Again, in a physics department, informal pressures were applied to an experimental physicist in an offbeat area:

> He likes music and has gotten interested in what makes woodwind instruments work. Everybody else in the department regards this as a very offbeat sort of thing, in the past of physics, but he differs with us. . . . I tried to dissuade him from this, and so did some of the other members of the department. But this is what he wants to do, so if we are unable to dissuade him we drop the subject.

Most of the examples cited in the preceding paragraphs involve "discrimination" against those in deviant specialties by departments in which the specialty has either weak representation or none at all. When the deviant specialty is well represented in a department it can sometimes act as a veto-group, insisting on getting its fair share of positions.[34] In the absence of consensus, "fair share" tends to be traditionally defined in terms of precedent. Many departments having deviant specialties use a quota system for appointments. For example, the chairman of a large mathematics department said:

> This department is run very democratically. All policies and appointments are decided by the tenure staff. . . . We are expanding because of increased enrollment. . . . We want roughly to cover all areas and set down percentages of people expected in that area. We tried to get outstanding men in each area . . . and then expand with younger men. . . . *Can the plan be changed easily?* You would have to convince the department. There would be an outcry if it meant expanding one field at the expense of others. People are not exactly jealous, but they do like to have those with whom they can talk, and a man in analysis can't talk to one in foundations. . . . As it is, some people in this department can't be counted in one area, their fields

overlap so much; they are split up in counting the number in each area. This kind of thing is just a rough guide.

A deviant specialty well-represented in a department can often maintain its position. Thus, while there may be relatively few members in the deviant specialty within the entire discipline, they may be concentrated in a few institutions. The establishment of this form of quota system often means that the representatives of the deviant specialty in the department are given autonomy in deciding on new appointments in their area. For reasons to be considered later, such groups tend to form around eminent scientists who are also forceful leaders, hence autonomy for the deviant specialty means autonomy for the full professor in that specialty and subservience to him by junior faculty in it. Many examples of this could be cited; one informant from a chemistry department said:

> Here I have been practically dictating biochemistry for the last thirty years. There wasn't anybody else so I had to do it, until in the last ten years we have had three or four more. At first the other chemists thought that biochemistry was rather pot-washing stuff, but I made them eat their words over the years. So they didn't give a damn what we did.

Specialties within university departments also dispute other such matters as access to students, curricula, and the distribution of laboratory facilities. The last of these appears not to have been a matter of contention in any of the departments in which interviews were conducted. Scientists in leading U.S. university departments obtain research grants from outside of the department, and they seldom have serious difficulty in obtaining small-scale grants. (Large capital investments are something else again and will be considered elsewhere.) In other places and at other times the situation may have been different. It has been alleged that the newer areas of biology—microbiology, molecular biology, etc.—have not developed strength on the European Continent because the established biological disciplines can effectively prevent the newer and "deviant" specialties from obtaining positions, research resources, and advanced students.[35]

The organization of the teaching curriculum is another possible focus for disputes between specialties. (Of course, with regard to curriculum, appointments, and other problems facing university departments, many other factors may give rise to disputes, and the disputes themselves are often conducted at very low temperatures and may hardly deserve the term "dispute." C. P. Snow's novels, especially *The Masters*,[36] although concerned with a form of organization rather different from the U.S. university department, show how various factors combine in the formation of collective decisions about personnel in a Cambridge college; as his work suggests, the most intense hostility may be aroused by objectively trivial issues.) In the physical sciences, the instruction of undergraduates is not likely to cause serious conflict between specialties. Curricula are more or less the same throughout the nation, and each professor has the academic freedom to teach his own courses as he sees fit. In seriously split departments, problems may arise; for example, a mathematician said:

> [This] mathematics department is not typical. . . . X university is quite different. It is much more homogeneous than [this one], which is very heterogeneous. It is difficult to get agreement here. This is not only the result of mathematical differences. For example, on the curriculum: one man may want his course earlier, with no prerequisites, in order to get more students. This kind of empire-building goes on. And it is hard to get any change in the department, it is so large. If a problem is posed, many different solutions will be presented and as a result no change will come about.

With the graduate curriculum, more serious problems may arise. Students in deviant specialties may be required to take courses and be examined in what seem to them irrelevant subjects, and this may threaten the training program in the deviant specialty. A graduate student in mathematical logic said:

> Many students in logic feel imposed upon by the requirement that they take mathematics courses in which they feel little interest—and, similarly, logic students in philosophy might feel the same way.

A distinguished logician who teaches in a philosophy department spoke similarly:

> [At X University] the professors of mathematics were not really interested in foundations problems. . . . A few of their students somehow became interested, by reading this or that, and so when I gave from time to time a course in the logical foundations of mathematics, a few of them came over; sometimes I would have from one-third to one-half students from mathematics in such a seminar. When they took this course I told them they also needed some logic, and many took some. . . . Then the problem arose when they asked to write a Ph.D. thesis with me, and to my regret I had to tell them it was practically impossible. Of course they could write with me, but it must be in philosophy, which meant they had to learn the whole of the history of philosophy, the whole of metaphysics, the whole of value theory, ethics, and aesthetics; but the students had studied in mathematics and were interested in it. . . . The mathematics department would count only a small number of nonmathematics courses toward completion of the degree in mathematics. Some professors discouraged students from taking philosophy courses as being of little value for work in mathematics.

Logic, however, is a reform specialty; logicians usually insist that their subject is part of mathematics, and they tend to encourage their students to take mathematics courses. One logician, on being told that students in logic sometimes objected to taking mathematics courses, said:

Students always feel this way. But if you're going to study the foundations of mathematics you must know something about it. So I don't believe they take more mathematics than is necessary. It depends on what you'll do when you have the degree; if you are to teach mathematics, you must learn it.

In rebellious specialties, professors are more likely to take the side of their students against requirements that they take courses in the central areas of the discipline. Many universities have ways in which intradepartmental conflict on such matters can be circumvented; these are interdepartmental committees for a marginal specialty, group majors, and so forth. A molecular biologist serving in a zoology department said:

At [this university] there is a strong tendency in departments such as zoology to hang onto tradition; this is done under the banner of "breadth." So the graduate curriculum requires much in totally unrelated branches of biology. . . . Evasion is possible. One of my students who objects to what they call "bird-watching" can take a Ph.D. in a group curriculum, for example in biophysics, and do his work with me. . . . This is fairly common. I always have one or two students like this. It is more common in other departments. X in physiology always has all his students in a group curriculum.

This procedure amounts essentially to granting the individual professor considerable autonomy in the instruction of his own students.[37] The degree of autonomy appears to be even greater at universities where the classical departments of biology no longer exist—zoology, botany, and so forth—and all biologists serve together in a biology "division." The procedure obviates the possibility of intradepartmental conflict over graduate curriculums and standards, although it has the obvious disadvantage of making decisions about new appointments even more difficult, since a division is a far more heterogeneous group than is a department. While instructional autonomy reduces one source of strain, it has a number of disadvantages. It makes the graduate student far more dependent on a single professor, and in the U.S. system, where professors often move between universities but students seldom do and where professors often take leaves of absence, the student may be left stranded. From the point of view of the department and the discipline, it increases the possibility of relaxation of standards for instruction and dissertations. Such autonomy is most likely to be given in elite institutions, where the faculty have high standards, so that in practice the threat is not severe. In universities of lower prestige, professors will have less autonomy in dealing with their graduate students; this permits the collective maintenance of standards but increases the strains between specialties with regard to them.

Granting instructional autonomy to professors can be considered a *primary adaptation* to interspecialty strains. It permits specialists with rather different goals to coexist in harmony in the same departments, without instituting any kind of structural or institutional change. Another primary adaptation is the already mentioned practice of making new appointments on a quota system for the different specialties. Such adaptations are consistent with the antiorganizational, anticonflict sentiments among most scientists, for example, the molecular biologist already quoted. Although he noted the differences between himself and his zoology department colleagues, he said that he personally desired to avoid conflict:

> There are different attitudes regarding which way [departmental] expansion should go. . . . The competition I'm familiar with is very gentle. There are no fights, we try to reach agreement. I don't know how it would come out in a fight. . . . I have a tendency to avoid conflict. Conflict is just as likely to have negative results as positive ones. . . . The result depends on how much conscious effort people make to get along.

Compare the remarks of an eminent logician:

> [E. G. Boring's] story of the history of the Harvard psychology department was interesting, how they had to struggle with the philosophers. . . . They had to struggle for years to convince the philosophers and the administration that not everybody knows psychology, that it's a science. . . . I said to Boring that there should be a new department of the logic and methodology of science. . . . Would your university be willing to make such a thing? Harvard was leading in many fields. He said, "If you had gone through the years of struggle to make a new department of psychology, you wouldn't do it for a second time. I'm all in sympathy with your ideas, but make your own struggles wherever you are." I said, "I'm not a man of struggle. I write books and I think about problems, and I'll leave it to other people to do it."

Nevertheless, such primary adaptations have disadvantages; they make departmental decision-making more rigid, they increase the dependence of graduate students on professors, and they lead to the relaxation of standards for graduate students.

SCIENTIFIC SOCIETIES AND JOURNALS

Organized scientific societies are primarily responsible for the maintenance of communications channels in science, for the conduct of scientific meetings, and for the publication of journals. Journals are also published by individual scientists and research establishments, although they, too, have representatives of the larger discipline on their boards of editors. Problems often arise in distributing access to communication channels. For example, the major journals always have backlogs of unpublished articles, and only a few subjects can be treated in special symposia at society meetings. Societies delegate responsibility for allocating access to communications channels to boards of editors and program chairmen of society meetings. There may often be competition between specialties for space and time, yet in general serious complaints about this are not raised; when goal consensus and a legitimate hierarchy of specialty prestige exist, a system of priorities can be instituted that will be accepted by most scientists.

Michael Polanyi has described how such a system might operate under typical conditions; his experience is largely in Britain, where disciplinary barriers are less important and the role of national informal leaders is more clear-cut; still, it probably applies to science in other countries as well.

> Scientific opinion exercises its power largely informally, but partly also by the use of organized machinery. . . . Even within the fields recognized at any particular time, scientific papers can be published only with the preliminary approval of two or three independent referees, called in as advisers by the editor of the journal. The referees express an opinion particularly on two points: whether the claims of the papers are sufficiently well substantiated and whether it possesses a sufficient degree of scientific interest to be worth publishing. Both characteristics are assessed by conventional standards which change with the passage of time according to the variations of scientific opinion. . . .[38]

When there is no goal consensus, or when many scientists deny the legitimacy of the prestige hierarchy of disciplines, the actions of editors and their referees will be challenged; specialists in deviant specialties may question the *right* of such formal authorities to restrict their access to channels of communication. Members of deviant specialties frequently complain that their contributions are denied enough space in existing journals. For example, a leader in the movement to establish an Association for Symbolic Logic twenty-five years ago said:

> The plain fact was that neither philosophers nor mathematicians were giving proper recognition to the field. There was active opposition in both quarters. . . . The opposition wasn't really of the violent sort, it was just unwillingness to give as much space in the journals and time in the meetings and appointments to individuals in this field as some of us then thought it deserved. Probably there were some diehards who were really violently opposed; they wouldn't speak to me, but from various overtones I gathered that some people were really pretty violently opposed.

The program chairman for an annual meeting of the National Biophysical Society described some of the tensions in that group:

> There are two main classes, the molecular biologists and the more physiologically oriented ones. There is a group more clinically interested, but they're not very well represented in the Society. They are more likely to be

in a health physics society; I think there is also a medical electronics group. And then there is a small group of more theoretically oriented biophysicists. . . . We're quite democratic, which is a way of saying we have almost no qualifications. Anybody who has any interest can join. *Is there any tension in the society between those who are physiologically oriented and those oriented to molecular biology?* Yes, there definitely is. While we don't talk about it, it's there and shows up in the meetings. I get this as program chairman. I try to keep some balance. *I suppose this shows up in the journal too.* Not as much, because at a meeting you have only so much time. *You can always schedule more sections.* We have some symposia which presumably interest all members. We have plenary sessions to discuss these things. It is mostly left up to the program chairman. . . . Soon the Council will meet to look at the plans for the meeting, and at that time specialties which feel slighted will make themselves heard.

The scientific society is similar to the university department in making certain primary adaptations to the internal strains resulting from goal conflict. As in the university department, these consist in "giving in a little." Deviant specialties may be given representation in the leadership of the society and on boards of editors of journals, sessions at society meetings may be devoted to them, and sections in the society may be formed of them—although the last named is a step of differentiation that could lead to still further alienation of the deviant group. A certain amount of space in journals may be devoted to the deviant specialty despite the editor's conviction that the topic does not "really" belong in it. The primary adaptations serve to alleviate the strains without recourse to changes in the structure of the discipline. However, they have disadvantages; collective actions by the discipline become more difficult because of the rigidity imposed by *de facto* compromises between groups that are not committed to common goals. Deviations from other scientific norms may become more likely. The standards of scientific journals, for example, will be relaxed when the decisions of editors and referees are based on "political" considerations instead of on standards to which they are deeply committed. Dissensus weakens social control.

GRANT-GIVING AGENCIES

Decisions by grant-giving agencies strongly affect the chances for survival of deviant specialties, and they may have considerable effect on the relative emphasis given to other specialties by scientists. These agencies are often directed by scientists, and they also rely heavily on the advice given them by scientists serving on advisory committees or study sections. Advisory committees are established for more or less extensive areas within disciplines or cutting across disciplines; they usually include scientists with a variety of interests.

Several of my informants had served on such committees. My initial hypothesis was that strains between scientific specialties would be expressed in the committees in much the same way they are expressed in university departments or the councils of scientific societies. That is, as long as goal consensus exists and specialties can be ranked according to generally accepted estimates of their importance, decisions are readily made. The ideal standards expressed by one of my informants could then be approached:

> I think only in terms of the reasonableness of the application and the evidence of the competence of the investigator. I don't think it makes any difference what the project is as long as it is a good project and the investigator is competent to carry it through.

When serious cleavages exist in a discipline, however, when scientists are unable to agree on the relative value of the work in different specialties, one could expect to find evidence of conflict between representatives of different specialties on these committees. No informant described any such disagreements.

Most agencies are sensitive to such possibilities and take quick action to lessen the effects of disagreements. A biochemist who had worked with both the American Cancer Society and the U.S. Public Health Service said:

> Where the tendency to funnel funds into narrow areas exists on committees, the Public Health Service tends immediately to set up a separate study section for this type of concentration. Just last summer this happened with respect to the area of nutrition. The PHS had some complaints that grants for nutrition may not have been receiving the favorable consideration the applicants thought was merited. The biochemistry section was saying that this was too applied, it wasn't biochemistry, it wasn't for us to handle. I was a member of an *ad hoc* group to look into these particular types of grants to see if there was merit in [the complaints]. . . . *When a new section is set up, is this usually the result of some dissatisfaction with existing procedures?* Yes, the problem is recognized. *Is this usually felt first by members of a section or by those who apply for grants?* I think it is recognized first by—those who are in the sections might see it developing—but it is first recognized when the PHS has before it on the table the recommendations of their many study sections. It is a matter of counting up areas that are approved, areas that may be disapproved, and areas where there may be uncertainty. I think it develops very naturally from an administrative viewpoint.

An immunological chemist who had also worked with the Public Health Service–National Institutes of Health confirmed this opinion.

This behavior by grant-giving agencies can best be understood by considering their own goals and distinctive problems. Colvard has shown that the nonprofit foundations that support basic research need to justify themselves and do so partly by demonstrating that they can support new types of research more effectively than other agencies—that they are inherently flexible and able to allocate their resources quickly to promising new fields of investigation. To achieve this flexibility the foundations must maintain their autonomy, but they

> find themselves continually vulnerable to control by groups in their environment. The groups include not only legal agencies and political groups but also—and of most interest here—two functionally related kinds of professional groups: (1) the professions themselves, especially those concerned with scientific research, and (2) professionally oriented, but broader organizations such as universities. Although it is the major goal of risk-capital foundations to support and stimulate highly promising innovative research and experimentation for which these two groups cannot obtain conventional support, many of the administrative procedures used in dealing with them must be understood in part as organizational defenses of foundation autonomy. More specifically, we contend that staff specialization, expert consultation, executive investigation, cost participation, and program and client concentration—common administrative procedures used in association with the project method—are all partly meant to reduce the foundations' vulnerability to control by professional groups.[39]

The same devices are used by governmental agencies that support basic research. There are many grant-giving agencies, governmental as well as nonprofit foundations, and they must compete with one another—indirectly for funds, more directly for public support and for the opportunity to contribute to eminent scientists. Public support is claimed on the grounds that the agencies can aid and stimulate major areas of innovation, areas which may ultimately be of practical utility, and thus no agency can afford to become strongly attached to a limited range of scientific goals.[40]

The competition between grant-giving agencies, their autonomy, and the freedom of action retained by scientists, all serve to increase the flexibility of the disciplinary organization of science. Specialties that encounter resistance within their disciplines may nevertheless receive support from outside. In countries and periods in which the degree of support for research has been far less, or where, as in the Soviet Union, the same degree of competition between agencies does not exist, financial aid to new areas of investigation may not be obtained as readily[41] in the face of resistance from traditional areas.

PRIMARY ADAPTATION TO CONFLICT

Goal conflicts result in disorganization. The growth of a scientific specialty having members who pursue goals thought to be inappropriate in their discipline, i.e., a deviant specialty, tends to produce other types of deviation. The most serious of these is deviation from norms of independence; formal pressures are exerted on those in the deviant specialty to induce them to select types of research problems felt to be more appropriate in the discipline. When formal pressures are used this way, the system of control through the award of recognition is weakened or abandoned. The distinctive feature of the award of recognition as a means of control is that it serves to emphasize the importance of individual commitment to higher norms and values; formal pressures are less likely to support such commitment.

The three most common areas of strain between a deviant specialty and the remainder of the discipline in which it is included occur with respect to university appointments, the instruction of graduate students, and opportunities to publish. In each of these areas, formal pressures may be used; for example, members of deviant specialties may be denied appointments. However, scientists are reluctant to apply formal controls. Instead, they adapt principles in various ways in order to control the strains produced by goal conflict. These adaptations may be called "primary," since they do not involve structural change. Three types of primary adaptations have been discussed: allocating appointments on a quota system, autonomy in the instruction of graduate students, and allocating space in journals on a quota system.

These adaptations may be sufficient to control strains until their causes have been overcome. After a period of time the incompatibility of the goals of the deviant specialty and the larger discipline may be perceived as only apparent. In the recent past chemists interested in quantum mechanics have been able to show that their work is "really" part of chemistry and not just physics, and metallurgists interested in the crystal structure of solids have been able to demonstrate that this is "really" metallurgy and not physics or chemistry; similarly, in the future mathematicians may come to believe that the general goals of mathematical logic are the same as those of mathematics. Alternatively, the deviant specialty may prove to be a blind alley and interest in it may die out, or it may become a minor interest in some other specialty. Thus, "biometrics" in the sense that it was cultivated by Karl Pearson and W. F. R. Weldon is no longer a distinctive specialty in the biological sciences.

On the other hand, the perceived incompatibility between the goals of the deviant specialty and the larger discipline may not decline but rather increase. If so, the disadvantages of the primary adaptations may be perceived as serious problems requiring structural changes for their resolution—secondary adaptations. Primary adaptations involving establishing "quotas" for appointments in departments or the inclusion of articles in journals directly subvert norms of excellence, for they mean that the competence of an individual or the quality of his research is considered less important than his research specialty. As Merton says of rebellion in the anomic situation:

> It presupposes alienation from reigning goals and standards. These come to be regarded as purely "arbitrary." And the arbitrary is precisely that which can neither exact allegiance nor possess legitimacy, for it might as well be otherwise.[42]

Furthermore, primary adaptations that involve granting intructional autonomy in the deviant specialty lead to greater dependence of graduate students on a single professor and may lead to the relaxation of standards.

The primary adaptations involve the withdrawal or segregation of the deviant specialty from the system of recognition and prestige-ranking of the larger discipline. But this isolation, resulting from the alienation of specialists from the goals of the larger discipline, may then lead to more intense conflict and, eventually, to the formal differentiation of the discipline.

Disciplinary Differentiation

Structural differentiation re-establishes social control. It begins with the organizational controls called into play when deviant specialties challenge the legitimacy of the informal organization of a discipline. It results in the formation of a new discipline with its own organizational controls—university departments, scientific societies, and channels of communication. After differentiation takes place, the organizational controls are consistent with the prestige hierarchy of specialties in the discipline; as a result the controls of formal organization are less likely to be used.

Many characteristics of scientists and scientific organizations inhibit differentiation. Scientists deprecate concern with organizational affairs; for example, "departmentalism" is a favorite whipping boy,[43] and many scientists tend to agree with the remark, "I really think that this [scientific] society business is of very little use to science." When formal organization is important, the value of information or suggestions depends on the status of the person who offers them, and this conception is alien to science.[44] Scientists rely on rational arguments based on empirical data available to all qualified observers, and this conflict of ideas is basically different from organizational conflict, where arguments are based on precedent, authority, and market value. Most scientists express a desire to avoid conflict. The sentiments already quoted are typical: "I have a tendency to avoid conflict," and "I'm not a man of struggle." The withdrawal of scientists in deviant specialties from administrative concerns and organizational conflict is facilitated by the primary adaptations of disciplinary organizations to goal conflict. Although these scientists are restricted collectively, they are granted freedom to pursue their work as individuals.

Scientists in deviant specialties are at first less likely to identify themselves with the emerging discipline than with the one to which they are formally affiliated. They will have been trained in a department of the traditional discipline, few of their departmental colleagues will be specialists in the deviant area, and they are usually affiliated with traditional scientific societies. Tendencies to identify with the traditional discipline are stronger when it has a higher prestige in the academic community than the deviant specialty; the status of statisticians as well as of applied mathematicians, for example, is enhanced if both are identified as mathematicians. Two of the statisticians interviewed for this study were employed in a mathematics department. Although they had received their doctorates in a department of mathematical statistics, taught only statistics courses, did research only in mathematical statistics, and published mostly in the *Annals of Mathematical Statistics*, they preferred to identify themselves as mathematicians rather than statisticians or mathematical statisticians. As long as individuals identify with the traditional discipline, organizational differentiation is unlikely to occur, even if the traditional discipline is under great internal stress.

Furthermore, at first, there will be no way of legitimating identification with a new discipline. In the absence of an ideology—a formula legitimating a distinct type of organization—claims for special treatment will be neither advanced nor recognized. Members of the deviant specialty will be only slightly aware of the history of their specialty and the distinctive services it can perform for science and the larger society.

Finally, differentiation will be inhibited because the organizations in which basic research is conducted tend to be rigid. Universities are organizations with a long history and well-established traditions. They are relatively well insulated from the rest of society and seldom need to change their structures in response to external pressures. Because the professional staff in universities is usually influential in determining policies, changes cannot be brought about by an administrator alone. If differentiation is to occur, then, a large group of professionals must be convinced of the desirability of the innovation; since the innovation is likely to be "permanent" it will not be accepted without careful consideration.

In summary: disciplinary differentiation requires leadership, men who are not reluctant to enter into organizational controversy; the development of an ideology that justifies claims on the wider scientific community and facilitates identification with the emerging discipline; and techniques for incorporating the new discipline into organizations that conduct research.

SCIENTIFIC PERIODICALS

Communication precedes community, and community precedes self-identification. Scientific publications devoted to a special field precede the emergence of the field as a discipline, and the emergence of the discipline precedes the identification of scientists with it. A central thesis of this work is that the primary reference group of the scientist is composed of those who read his published work. (Since good paper may last two millennia or more before decomposing, it is possible for scientists to have their reference groups many generations hence.) Today almost all work of importance in the exact sciences appears in scientific periodicals.[45] As long as the readers of the journals in which a scientist publishes identify themselves, and are identified, as members of a traditional discipline, he will find it difficult not to identify himself as a member of that discipline. Only when a periodical is established that is devoted to a field with its own distinct goals and standards will it be possible for him to conceive of himself as a new kind of specialist; only then will it be possible for a self-conscious community of specialists to arise.

The recognition sought for and awarded in periodical publications controls the scientist by forming his conception of who he is. If his readers are a select group, oriented to the goals and standards expressed implicitly in the new periodical, his contributing information to them and being recognized in print by them will contribute to his identification as one of them. (This is one way in which the strictly scientific publication differs from the text or the scientific popularization; the response by the audiences of the latter are valuable only with respect to the writer's conception of himself as a teacher. These audiences are not, or are not considered to be, capable of criticizing the substance of his work and therefore not capable of recognizing his originality.)

The formation of scientific periodicals is a critical step, then, in the differentiation of new disciplines and, necessarily, one of the first steps. Of course, there are many journals,[46] and it is not hard to establish new ones. To become effective as the medium of a new community, a new journal must have distinctive goals and high standards. The major threat to such efforts is that, to a new journal, writers send manuscripts already rejected by journals representing a traditional discipline. When this occurs—and it is probably the typical fate of publication efforts—a journal becomes known as one with lower standards but the same general goals as traditional journals, rather than one with high standards but different goals. As a result, the scientist conducting research remains oriented to the audience of the traditional journals.

Even journals that establish their distinctiveness must face criticism for having lower standards than the traditional ones. Such criticism is often justified. Since it has as yet no specialized educational program in its field, an emerging discipline must rely on the work of relatively untrained scientists. Thus, more than half a century after the event, Lancelot Hogben has been able to criticize the work of Karl Pearson. Pearson's mathematical contributions to regression theory were published under a biological title in the *Philosophical Transactions of the Royal Society*, Hogben writes:

> and on that account exempt from exposure to the scrutiny of mathematical colleagues sufficiently familiar with the theory of the combination of observations to recognize its abuse as well as its uses. . . . Being a skillful manipulative mathematician equipped with a vigorous command of the English language, he had no difficulty in recruiting a militant following to spread a gospel which handicapped the progress of experimental genetics in Britain for at least half a generation.[47]

Eventually the Royal Society editors rejected such work, allegedly because it was too mathematical for the biologists and too biological for the mathematicians.[48] However, Sir Francis Galton's private fortune made it possible to found *Biometrika*, with Pearson as editor and no possibility of rejection of his work by hostile referees.

In a similar way it has been alleged that Freud and his followers were able to publish as much as they did only because a wealthy benefactor willed them the money to found a publishing house. Ernest Jones went to considerable effort to disabuse us of this notion in his biography of Freud.[49]

Because of criticism based on the standards of traditional disciplines, adherents to a new discipline may find it difficult to maintain their self-esteem even if they come to conceive of themselves as belonging to a new kind of specialty. Possibly a kind of defense reaction could enable them to maintain their self-esteem by taking a strongly negative stance toward traditional disciplines: Karl Pearson could condemn the errors made by specialists in many traditional fields, Frege could be scornful of the ignorance of mathematicians about the number system, and graduate students in biophysics today might refer satirically, as some do, to "bird-watching" in the classical biological fields.[50]

Given specialized communications channels, specialists in an emerging discipline can withdraw from the traditional disciplines. They can and do develop a new and distinct terminology, which further serves to isolate them from what went before. The results may be a self-conscious group alienated from the existing disciplinary organization of science.

IDEOLOGY AND UTOPIA

Every established discipline possesses an ideology, a more or less explicit justification of its privileges and the claims it makes upon the scientific world and the larger society. These ideologies are partly alleged facts about the contributions of the discipline and partly evaluations about what is or should be considered "interesting" and "intrinsically important." Established disciplines are well articulated with groups and organizations in their environments, and their ideologies are restricted in scope and oriented to specific audiences, primarily within science and scientific organizations.

A scientific ideology has various facets, each corresponding to the audiences to which it is addressed and the functions it performs. One aspect is concerned with the jurisdiction of the discipline, with what it includes and excludes. Since the jurisdiction of the discipline is directly related to the rewards and facilities it can claim—for example, the course load of a university department will affect the number of faculty positions available to it—jurisdictional disputes may arise between disciplines. (I expected this to be relatively important and asked my informants several questions about it, but in the universities and the disciplines within which I interviewed, such disputes are apparently rare and unimportant.) Thus one function of the ideology of chemistry will be to mark its jurisdictional lines with physics, the life sciences, and the engineering disciplines. However, there is no general tendency for a discipline to be imperialistic. Another function of its ideology is to enable the discipline to resist claims made on it by those for whom it has instrumental value. The "pure" aspects of the science will be stressed in this regard—its intrinsic interest and general importance, as opposed to its particular utility for other types of scientists.

Disciplinary ideologies also have internal functions. As statements and justifications of goals, they regulate relations among specialties within the discipline, and by contributing to the self-conceptions and self-esteem of specialists they help maintain its solidarity. Well-established disciplines have little need of explicit ideological formulations, and their ideologies are an unsystematic collection of parables, heroic myths, and invidious distinctions between specialists and nonspecialists. Chemists, for example, assert that physicists leave the "really hard" physical-chemical problems to them, that life scientists do not understand the chemistry of the objects with which they deal, and that chemical engineers more often resemble cooks with recipes than theoretical scientists. While scientists usually have little interest in the history of science, each discipline possesses a kind of historical myth, one emphasizing the unity of the discipline and its continuity in time.[51] This historical myth is presented in short historical notes in textbooks and in such quasi-popular works as Eric Temple Bell's *Men of Mathematics*, a book almost every graduate student in mathematics reads sooner or later.[52]

Corresponding to the ideologies of established disciplines are the utopias of newly emerging disciplines, justifications of proposed changes in the structure of science whereby the new discipline will gain a more secure position. Disciplinary ideologies tend to be restricted in scope, oriented to specific audiences, and implicit, while disciplinary utopias tend to be "imperialistic," almost unrestricted in scope, oriented to very general audiences, and explicit.

An emerging discipline is vulnerable and seeks assistance from many sources, outside of the scientific community as well as within it. Explicit claims are made about the potential utility of the field for all or almost all other fields, both pure and applied. The utopia may be "utopian" in the pejorative sense, by claiming to be the basis of understanding almost everything.[53] Consider, for example, the utopia of mathematical statistics. In its extreme forms, it includes the claim that all scientific propositions are statistical propositions, hence the statistician has something to contribute to all fields, from nuclear physics to literary criticism. Furthermore, it is claimed that statistics can be valuable to all applied fields, from medicine to gambling. Opinions like the following, ascribed to Harold Hotelling, may be extreme, but they are not especially unusual:

> In the dark days of World War II, at M.I.T. I heard him express in all seriousness the conviction that the Allies would surely win the conflict, because they had (in Britain, the United States, and India) the leading masters of modern statistical theory in contrast to Germany and Italy which had none. And I believe it was Hotelling who during the War suggested *Econometrica* be discontinued lest its contents lend aid to the Enemy's war effort! . . . But the history of ideas shows that strong faith is the pioneer's greatest asset.[54]

Karl Pearson put forward a version of this utopia in a series of influential books and essays, particularly *The Grammar of Science*, as other leaders in the field have also done.[55] (Often the claims relate not only to the "discipline" in general but to a specific theory advanced by its leaders. Pearson's claims, for instance, were closely related to his own theory of moments and Galton's genetic theory.)

Other emerging disciplines have similar utopias. Perhaps most serious specialists neither make nor accept these claims—they are most likely to be made and accepted by the leaders of the emerging field, its popularizers, and nonspecialists who hope to obtain useful knowledge. But this does not mean such utopias are unimportant—their claims serve mainly to attract support from members of the wider scientific and lay public. This is an important function, and it may be asserted tentatively that *every* science has advanced a utopia at the stage when it was striving to become established. Recent examples are the fields of genetics, psychoanalysis, and small-group theory. In the early part of the nineteenth century, at a time when chemistry was advancing to a disciplinary status on a par with physics, Humphry Davy made a speech in which he pointed out the great importance of chemistry for such fields as mechanics, natural history, mineralogy, botany, zoology, medicine, psychology, astronomy, and the practical arts.[56] It is dangerous to carry the idea of a "disciplinary" ideology back to a time before disciplines in the modern sense existed; even so, interesting parallels can be found. At first it is difficult to recognize the utopia of classical physics, since it was so completely victorious; the world was alleged to be *nothing but* "matter in motion"; physics studied "primary qualities," whereas other disciplines were concerned with mere "secondary qualities." One must go still farther back to find a mathematical utopia: modern mathematicians may *say* "God is a mathematician" (or a geometrician, or an arithmetician, depending on the speaker's special field), but the Pythagoreans really believed it: musical tones, the movement of heavenly bodies, personality, and everything else—all were thought to be governed by number and mathematical relations.

To complete its utopia, an emerging discipline must create a history for itself—it must discover its past.[57] Those who contribute to emerging disciplines do so at first in ignorance of one another's efforts. Using different terminologies and perhaps studying slightly different types of phenomena, they cannot understand one another and do not recognize one another as colleagues.

Sociologists will recognize this as the situation prevailing in sociology around the turn of the century—such men as Max Weber and Émile Durkheim were essentially unaware of each other, and most American and English sociologists were scarcely aware of either of them. Sociology had largely *national* traditions. As sociology has emerged as a discipline, attempts have been made to create a history transcending national boundaries and the boundaries of specialized "schools."

The same has been true in other emerging disciplines. Freudenthal has shown that logicians investigating the foundations of mathematics in the latter part of the nineteenth century were often aware of the work of their fellow nationals only, and not always even of them.[58] Russell and Whitehead, Peano, Hilbert, and Frege were initially unaware of one another. In the years following 1900 they discovered one another, but it took many years before logicians generally came to feel that they shared a common history. The search for a common history was best manifested by the decision of the founders of the Association for Symbolic Logic to devote most of the first volume of their *Journal* to an annotated bibliography of works in symbolic logic from the time of Leibniz to the nineteen thirties. Many of the founders of statistics as a discipline had a similar interest in the history of their field. Karl Pearson, for example, spent many years writing his monumental *Life and Letters of Francis Galton*, as well as other historical investigations.

Even as a "science which hesitates to forget its founders is lost,"[59] a discipline which ignores its founders is stillborn. Socially vulnerable disciplines seek legitimacy and a basis for group identification. One means to this end is the discovery of a common history—a history that is *made* common. This "history" may be largely mythical; all the same, its social effects may be considerable. Less vulnerable disciplines, of which molecular biology may be an example, have less need for such historical myths. Specialists may be aware of sharing a common tradition and contributing to it without feeling the need to write history.

Emerging disciplines require explicit utopias to *legitimate* their claims and to form the basis for the *identification* of scientists with the new disciplinary community. The legitimating function is relevant primarily for nonspecialists. The emerging discipline makes claims for special treatment which cannot be justified by the existing organization of science; scientists generally feel that the emerging discipline is and should remain a part of existing disciplines. The disciplinary utopia stresses the emerging discipline's distinctive aspects,[60] as well as its homogeneity, and states that the differences within it are small compared to the differences between it and other disciplines. In fact, however, this claim is likely to be unjustified.

THE IMPORTANCE OF MARGINALITY

Emerging disciplines are inherently heterogeneous. Specialists lack common traditions and education and come to the new discipline with different techniques, different terminologies, and different substantive interests; they share some goals and a confidence in the importance of them. The emerging discipline cannot restrict its membership: to advance its claims it must seek support from many sources.

Usually scientists with applied interests are thrown together with those whose interests are purely theoretical. As Ben-David has suggested from his study of bacteriology and psychoanalysis, this may be partly because basic innovations are most likely to be made by "role hybrids," those with both pure and applied interests—such men as Pasteur, Koch, and Freud.[61] This has happened not only in the biological sciences with practical applications[62] but also in economics[63] and statistics. On the other hand, an emerging discipline may contain only scientists with purely theoretical interests, although still from widely different backgrounds. Examples may include mathematical logic, marginal between philosophy and mathematics, and chemical physics, marginal between chemistry and physics.

The presence of such heterogeneity makes it easier to justify the claim that the emerging discipline does not belong in existing disciplines. Structural changes in disciplinary organization do not seem to occur unless the emerging discipline is marginal to at least two others already in existence. For example, the discipline of chemistry is highly differentiated, and the differences between physical chemists and organic chemists are often greater than the differences between the former and physicists and the latter and molecular biologists. Yet chemistry has never been formally divided into several disciplines, although some structural changes have taken place. In general, differentiation will be incomplete in the absence of marginality.

Heterogeneity produces internal strains within an emerging discipline. These strains may be concealed by a generally accepted utopia while the discipline is struggling for organizational rights. Later on they may result in internal disputes. For example, the President of the American Statistical Association in 1958, Walter Hoadley, devoted his presidential address to a discussion of "cleavages" within statistics, primarily the cleavage between mathematical statisticians and applied statisticians:

> [T]his cleavage often becomes so deep-seated that many statisticians find themselves literally forced to take sides, i.e., to classify themselves as members of either the higher-mathematical or non-higher-mathematical camp. The final result is often withdrawal of individuals and groups from general activities and discussions at national, regional, and chapter meetings into tightly knit groups or organizations evidencing pride in their command of, or contempt for, mathematics. . . . Certain mathematical statisticians may tend to give the impression that they now have a superior standing in the profession since they have a greater command of the latest statistical developments which admittedly seem to involve a great deal of mathematics.[64]

The final result of such disputes is usually the "purification" of the discipline qua discipline; those whose primary interests are applied or clinical —for example, psychometrics in statistics or nutrition in biochemistry— are excluded from university departments of the discipline, and separate societies for those with pure or applied interests are formed. In this manner, the Institute of Mathematical Statistics separated itself from the American Statistical Association, in 1935. The larger society may continue as a kind of federation of groups with divergent interests.

This later differentiation—the purification of the discipline—is accompanied by the transformation of the disciplinary utopia into an ideology. That is, expansive claims are less likely to be made, and the ideology may be more defensive than imperialistic; having achieved their goals, specialists in the pure aspects of the discipline begin to defend themselves against those who make claims on them. This transformation is likely to be resisted by those most strongly committed to the disciplinary utopia, the leaders of the emerging discipline.

LEADERSHIP

Most scientists avoid administration and administrative controversy. But the differentiation of a discipline inevitably involves both, and emerging disciplines tend to be especially dependent on those willing to assume leadership. These men may be disparaged as "empire-builders," but to be successful they must have won the respect of scientists in other areas through their discoveries. They may feel that they are defending not an organization but rather the truth as embodied in a theory to which they are strongly committed. The controversies that engaged men like Freud and Karl Pearson were largely substantive, although their actions resulted in the organization of scientific groups with distinctive goals.

The leader seeks for and obtains power. His power may be based on both the charismatic qualities attributed to him by his followers and the position he holds in scientific organizations. He may, as did Pearson and Freud, play a large part in formulating the disciplinary utopia; and he may symbolize the discipline to his followers. Frequently specialists in the area will have been either his students or in close personal contact with him in other ways.[65] He will demand loyalty and claim their support when controversy occurs. His authoritarianism may generate strains between him and his followers.[66]

The leader also obtains purely organizational power. He will seek the aid of powerful persons in the academic community or the larger community, and he will use his position as organizational leverage for obtaining autonomy for his specialty in research organizations. Thus, Karl Pearson, with the aid of the wealth of Sir Francis Galton, was able to establish the Galton Chair of Eugenics at the University of London.

STRUCTURAL CHANGE IN UNIVERSITIES

In the United States a well-established scientific discipline requires university departments consigned to it. Specialized journals and scientific societies may help maintain the distinctive goals of the field, but they are highly vulnerable. The heterogeneity of the societies makes it difficult for them to form a community that can more or less monopolize recognition for work in a special area. Since they can control neither recruitment into the field nor the socialization of recruits, their goals and effectiveness may change as their composition changes.

Looking at it another way, we may say that a department manifests a strong commitment on the part of the university. Departments reproduce themselves by providing relatively common training for recruits stressing the distinctive goals and standards of the field. The professorial members of departments cannot be easily dismissed by the university. Departments may be abandoned (or "reorganized"), but this is difficult and seldom occurs. Once universities establish departments, they are committed to them and to their continuing support. The "conservatism" and "rigidity" of universities can therefore be viewed as unwillingness to commit themselves prematurely. It follows that those who seek to obtain autonomy for an emerging discipline must initially search for organizational formulas involving less of a commitment than the establishment of a department.

Such formulas may center on either research groups and facilities or courses of instruction, depending on the relative importance of different problems facing the emerging discipline. One of my informants was on a faculty committee to investigate the development of molecular biology on his campus. He ranked the reasons for establishing some special organizational form for this subject in the following order:

(1) People now doing research in the area are widely dispersed among a number of departments; this makes communication between them less likely. It would be desirable for them to be together so they could offer common seminars and other meetings.

(2) If they were organized together, they could get more research resources. Everyone can now get as much in the way of grants as he wants. But grants for buildings and large-scale equipment are not obtained as readily. A separate organization would be more likely to get such facilities.

(3) In an institute or department people could share equipment.

(4) Curriculum problems rank fourth and are very unimportant. For example, although students in chemistry may have to take courses in chemistry that are relatively insignificant for molecular biology, this is actually unimportant. Graduate requirements are very flexible in all departments. Professors in molecular biology already have as much autonomy as they desire in this respect.

When facilities and the opportunity to form research groups are considered the most important problems facing a field, specialists will attempt to have some form of research institute established. Such an organization will be in a better position to obtain expensive research facilities and research assistance. Obviously, large capital expenditures represent strong commitments by the university involved, even if an outside agency provides the initial capital. Universities have different ways of handling the problem, but the procedure recently instituted by the University of California illustrates the range of commitments involved.[67] The University distinguishes among "institutes," "centers," and "projects." (Some other forms, such as "museums," need not be considered here.) The University is least committed to projects: these are expected to be terminated at some fairly specific future date. It is most committed to institutes, which are expected to be "continuing." Nevertheless, all such research units are to be reviewed by a special committee appointed by the president at five-year intervals, and the committees are directed to "submit a report appraising the need for the continuation of the unit."

Such units can be abandoned far more readily than academic departments. (One informant was an eminent biological scientist now retired. One of his keenest disappointments was the decision by the university to dissolve a research unit he had founded.) Their staff consists of university faculty, who are members of existing departments, and of specialized research personnel, who are usually denied tenure. Although students may conduct dissertation research within such units, final authority over instruction and dissertation standards resides in academic departments.

Sometimes the problems most keenly felt by members of a deviant specialty will involve instruction. Initially the problem will be solved by the establishment of interdepartmental committees for the instruction of doctoral candidates. Committees of this kind often exist for such fields as molecular biology, chemical physics, geophysics, and logic. As the number of students and faculty involved in such work increases, efforts may be made to offer specialized courses at lower levels. Dissatisfaction with the "preparation" of advanced students will be voiced, and it will be argued that specialized undergraduate courses are also needed. Since these are usually given by established departments, the proposal implies the establishment of a new one.

Whatever the type of organizational unit proposed, it is usually deemed desirable that an eminent scientist be appointed as its head. Sometimes he may be on the scene and the prime mover for the innovation. Or he may be brought in after the decision to establish an institute or department is made. The informant previously cited said that the establishment of an institute of molecular biology on his campus

> would require bringing in an eminent outside man. Only in this way could the problems facing such an organization be solved. . . . An eminent man can secure the respect a new program needs. . . . For example, X university appointed Y [a Nobel Prize winner] to direct its [new department in one of the life sciences].

Any structural innovation short of the establishment of a department lacks the critical power of collegial bodies, namely, power over the appointment of permanent personnel, and granting this power of departmental status is the final stage of differentiation within a university. In some cases differentiation may be incomplete. Organic chemistry and physical chemistry, for example, while formally together in a single department, may be more or less independent of each other within it. In some universities the department is formally divided into "divisions," each of which is headed by a vice-chairman. In other universities the differentiation is informal:

> The principal cleavage is usually between organic chemists and others. The organic chemists might constitute from 40 to 60 per cent of a typical department. If the department includes biochemists, they will be united with the organic chemists. Here, the organic chemists have a separate [faculty] seminar. We also have an organic staff meeting following the seminar. Similarly, the organic chemistry staff decides which organic chemists will

be hired, and they usually take care of the undergraduate organic chemistry courses, and only those courses.

Structural changes are usually made first by leading universities and followed later by others. Of course, not all innovations are imitated:

> If every distinctive name for the field of award is counted, there are now well over 550 "fields" in which the doctorate is awarded by one or another institution (but only one institution in almost 400 of them). This figure is directly comparable to the 149 fields in 1916 to 1918. Actually, of course, the number of "real" fields is much smaller: depending on how a "field" is defined, one gets between 60 and 80—all the others are variants, offshoots, or combinations.[68]

Universities do respond to some types of intradepartmental strains by establishing new departments, yet it is clear that this is usually only a local matter; only a fraction of such strains are both general enough and severe enough to warrant the formation of new departments.

"Departments" are especially important in U.S. universities, and therefore the preceding remarks about disciplinary differentiation refer largely to the differentiation of university departments. Elsewhere, other organizations may be more important. In the Soviet Union, the locus of disciplines appears to be in the component units of the Academy of Sciences, which, however, engage in instruction of advanced students. In Europe generally, universities appear to be less flexible than in the U.S.; the organization of faculties resists change more than the organization by departments.[69] In Europe, the primary collegial group is the "faculty" of full professors, and this group plays an important role in determining examinations, appointments and structural changes. Within departments nonprofessors have far less independence than their American counterparts, but each professor has far greater freedom than his American counterpart in determining the course of instruction and research in his department. In the U.S., while the department is the primary collegial group, the faculty as a whole is far less so. Rather, university presidents have considerable authority, an authority extending to decisions to establish new research units. U.S. universities are more flexible structurally than European universities, but this may be compensated for by the greater freedom of European professors within a traditional form of organization.

INDUSTRIAL AND GOVERNMENTAL LABORATORIES

Industrial and governmental laboratories are formed to achieve specific applied goals and have no responsibility for protecting the integrity of the scientific disciplines represented on their staffs. However, the needs and wishes of their scientific staffs usually influence them to recognize scientific disciplines in their organization.

When the organization of work groups in laboratories is based on disciplines, they may be called "specialist groups." These tend to be homogeneous and more or less permanent. When the organization of work groups is based on the problems to be solved, they may be called "task groups," which tend to be heterogeneous and of limited duration.[70] In practice, various combinations of these modes of organization are made. Task groups may be more effective for applied research; scientists working in them become committed to the goals of the organization, scientists from different fields stimulate one another, and the co-ordination among different specialties occurs on a continuing basis. Nevertheless, organizations may prefer specialist groups for other, nontechnical reasons. Professional scientists are oriented to recognition by their colleagues, specifically their disciplinary colleagues. They dislike being evaluated by superiors representing other fields, and they may find it difficult to agree with co-workers from other specialties. To the extent that they do become disoriented from their disciplines and reoriented toward the heterogeneous task group, their specialized competence may suffer; they may be unable to "keep up" with developments or make contributions to the discipline. Specialist groups, although they may inhibit co-operation between disciplines and that kind of creativity arising from contact with different perspectives, are more likely to ensure that scientists will be strongly committed to maintaining their own professional competence.

Industrial and governmental laboratories typically *follow* universities in their organizational procedures. In the most practical work, such as engineering development, this is unnecessary; engineers do not have, nor do they need, the same kind of identification with their profession as scientists do.[71] But in the fundamental or exploratory work performed by industry, when particular end-products are not yet known, industry is more dependent on the specialized competence possessed by scientific communities and must respect those communities.

DIFFERENTIATION AND SOCIAL CONTROL

The establishment of a new discipline removes the strains that existed when its members were incorporated in other disciplines. The harmony and efficiency of the organizations in which pure research is conducted is increased. Formal controls are less likely to be exercised—university department chairmen, journal editors, and scientific society leaders need not exercise an apparently arbitrary authority. Instead, controls will be largely informal. The system in which information is exchanged for recognition can be effective, since colleagues in organizational units seek recognition from the same or similar audiences. Since a general consensus exists regarding the relative importance of specialties within the discipline, organizational decisions about the allocation of rewards and facilities to specialties can be made in an orderly and acceptable manner. Finally, the autonomy of the scientific community is strengthened with the cessation of claims to the larger community by the formerly deviant specialty. This autonomy is threatened when groups within science seek to have nonscientific groups interfere with the organization of science.

Specialization and Structural Change[72]

Structural change in science has been considered here primarily because it reveals the limits of informal colleague control. We can learn how formal controls are articulated with informal controls in stable situations by observing what happens in unstable situations. Strains latent in many situations become manifest when they lead to structural change. Thus, ordinarily, scientists who deviate from the goals of their disciplines receive less recognition; formal sanctions follow on this: they may find it difficult to obtain positions, to teach students as they see fit, or to communicate their findings. This process may be difficult to observe in stable situations; it is studied more easily when deviant groups of scientists reject the legitimacy of formal controls and address appeals for protection and autonomy outside the discipline. Within disciplines, goal conflict cannot be resolved in the same ways that other scientific disagreements are resolved. Structural change alleviates organizational strains and makes possible a system of social control based primarily on informal colleague relations.

Change itself is an important problem, and this chapter can be summarized by bringing together the propositions relevant to social change that have already been made. Segmentation begins with cultural change, the appearance of new goals in the scientific community. Of course, new goals do not spontaneously "appear": scientists actively seek them. Those who discover important problems upon which few others are engaged are less likely to be anticipated and more likely to be rewarded with recognition. Thus scientists tend to disperse themselves over the range of possible problems.[73] The behavior is analogous to competitive behavior in the animal world:

> But the struggle will almost invariably be most severe between the individuals of the same species, for they frequent the same districts, require the same food, and are exposed to the same dangers. In the case of varieties of the same species, the struggle will generally be almost equally severe, and we sometimes see the contest soon decided. . . .[74]

In many disciplines, dispersion to avoid competition takes place not only over the range of problems available but over the range of institutions, with the result that few identical specialists will be found in the same organizations. "Complete competitors cannot co-exist."[75]

Dispersion may lead to isolation, both geographical and social. Scientists working on the most unusual research problems, not being encouraged elsewhere, may be concentrated in a few research establishments. Social isolation results when pursuit of different goals leads to the development of different terminologies, techniques, and modes of organization. Eventually communication between specialties may be difficult and uncommon. Research in one area will have only remote effects on research in another, and programs of instruction may become differentiated; the social equivalent of crossbreeding will occur less frequently:

> Isolation, also, is an important element in the modification of species through natural selection. In a confined or isolated area, if not very large, the organic and inorganic conditions of life will generally be almost uniform; so that natural selection will tend to modify all the varying individuals of the same species in the same manner. Intercrossing with the inhabitants of the surrounding districts will, also, be thus prevented.[76]

Dispersion and isolation may encourage cultural differentiation, but here the biological metaphor breaks down. Differences between specialties may be viewed as deviance by members of specialties that are traditional or central to the discipline, and attempts may be made to sanction such deviance.

Initially, attempts may be made to establish conformity by the use of formal sanctions—with regard to appointments, instruction of students, and access to communication channels. The exercise of these sanctions tends to be implicit. Unsuccessful candidates for jobs may not be told specifically why they are not appointed, papers may be rejected by journals for vague reasons, and a professor may find his students failing because of their "incompetence" or their "attitudes." Whether or not the recipient of the sanction is informed of the standards used in judging him, the matter does not become public.

Further development of the deviant specialty leads to overt social conflict. Those likely to be sanctioned publicly question the legitimacy of the standards used. Thus goals and standards are made explicit, and scientists and others will be made aware of the conflict. At first, organizations may attempt to reduce the resulting strains by primary adjustments. These usually amount to a limited range of permissiveness for the allegedly deviant group; they will be permitted a maximum number of appointments, limited access to channels of communication, and a higher degree of autonomy in instructing their own advanced students. Primary adjustments may make it possible to cope with conflict, but sometimes specialties will continue to diverge from one another, and the disadvantages of primary adjustments may lead to continued dissatisfaction. Such organizations as university departments and scientific societies will be made more rigid, scientific standards usually expressed by the award of recognition may become relaxed, and programs of instruction may suffer. Dissatisfaction with these failings may stimulate formal differentiation of disciplines.

Such differentiation requires special communication channels, the development of a disciplinary utopia, and successful appeals outside of existing disciplines. Leadership of an unusual sort in science—leadership unwilling to shrink from organizational controversy—will be necessary if these steps are to be successful. The establishment of communication channels and the development of a utopia make it possible for scientists to identify with the emerging discipline and to claim legitimacy for their point of view when appealing to university bodies or groups in the larger society.

At first, organizations may respond to these claims by structural innovations to which they need not be committed, innovations which they may abandon. In the university setting, this means such things as research institutes and interdisciplinary teaching programs, but not departments of instruction. Later on, separate departments for the new discipline may be established. This represents an almost irreversible differentiation, for universities are strongly committed to departments of instruction, and, through their graduates, the departments have reproduced themselves and established ties with groups in the larger society that employ them.

Structural change, especially departmental differentiation, is unlikely to occur unless the emerging discipline is marginal to at least two existing disciplines. When the emerging discipline is confined to a single existing discipline, it is apparently difficult for it to legitimate appeals outside of the disciplinary structure, especially appeals for structural change.

This analysis of social change can be conveniently presented in the schematic form in Figure 1.[77]

Differentiation results in the re-establishment of disciplines and specialties as the basic communities in science. New disciplines have internal structures similar to those from which they have differentiated, and this internal organization is characterized by the relatively great importance of informal relations.

From the point of view of the larger scientific community, however, continuing structural change implies a qualitative change in the organization of science. Modern science is no longer like the science of the seventeenth and eighteenth centuries. There are no universal scholars, and interdependence between disciplines and even among specialties within disciplines is small. This degree of specialization has implications for the organization of science that are reflected in the current problems of organizing university instruction and in allocating research facilities among disciplines. It also has implications for the place of science in the larger culture. Science no longer presents a unitary picture of the world to the nonspecialist, and the place of the scientist as a cultural leader is

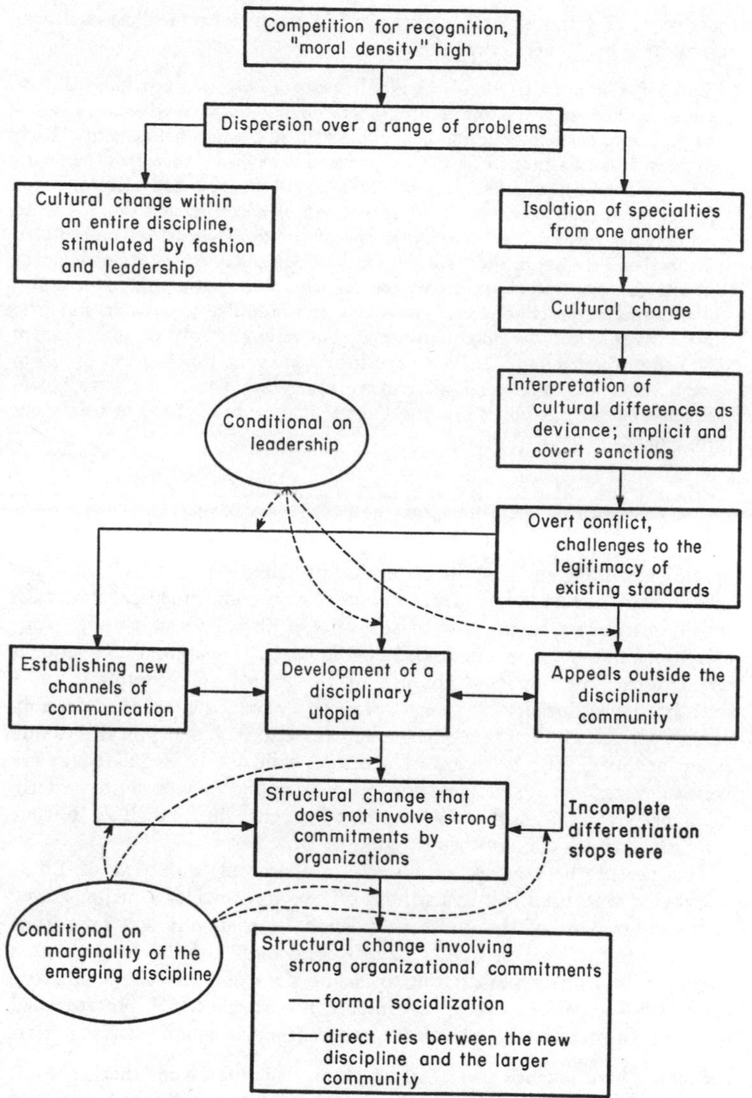

FIGURE 1. Differentiation in Science: Segmentation

thereby more dubious. This is expressed in a somewhat disillusioned passage written by Robert Oppenheimer:

> Today [as opposed to Plato's Greece], it is not only that our kings do not know mathematics, but our philosophers do not know mathematics and—to go a step further—our mathematicians do not know mathematics. Each of them knows a branch of the subject and they listen to each other with a fraternal and honest respect; and here and there you find a knitting together of the different fields of mathematical specialization. . . . We so refine what we think, we so change the meaning of words, we build up so distinctive a tradition, that scientific knowledge today is not an enrichment of the general culture. It is, on the contrary, the possession of countless highly specialized communities who love it, would like to share it, and who make some efforts to communicate it; but it is not part of the common human understanding. . . . We have in common the simple ways in which we have learned to live and talk and work together. Out of this have grown the specialized disciplines like the fingers of the hand, united in origin but no longer in contact.[78]

Specialization without Integration: Anomy

Disciplinary differentiation involves the specialization of individual scientists and the segmentation of disciplines. These complementary processes usually lead to the integration of scientists in disciplines, but under some conditions they need not have this consequence. If a branch of science is characterized by a *general* absence of the award of recognition, or if scientists in a specialty with low prestige believe they cannot feasibly change specialties, then recognition will cease to have its typical meaning as an incentive. The behavior of scientists will not be regulated in the way suggested here, and specialties in a discipline may have no orderly relations with one another. When this happens, the branch of science can be said to have an *anomic* division of labor.

This use of the concept of anomy is consistent with that of Émile Durkheim, who used it in his studies of law and suicide; a major source of his illustrations of the division of labor, both anomic and nonpathological, was the science of his day. Anomy was defined by him as a condition in which "the division of labor does not produce solidarity" because "the relations of the organs are not regulated."[79] He regarded the social sciences of the late nineteenth century as anomic communities:

> Scholars have installed themselves in them, some here, some there, according to their tastes. Scattered over this wide surface, they have remained until the present too remote from one another to feel all the ties which unite them.[80]

While the state of anomy has a variety of causes, the common element seems to be a "declassification" that casts persons into social roles and statuses in which their customary relations with others are broken and the standards formerly governing their behavior are no longer applicable. Among social scientists, this resulted from the rapid exploitation of new areas of investigation, while in the society at large it may result from the breakdown of traditional economic markets or the dissolution of families.[81]

Anomy is a form of social disorganization associated with social deviance (violations of norms). However, anomy as "normlessness" implies that social deviance is a minor component of the disorganization it involves; individuals may fail to conform to one another's expectations, but this is not so much deviance as the absence or ambiguity of norms. As Durkheim noted, individuals in anomic societies often feel exceptionally free and uninhibited: "There is no restraint upon aspirations."[82] People attach absolute value to their limited goals. In the sciences:

> What best manifests, perhaps, this absence of concert and unity is the theory, so prevalent, that each particular science has an absolute value, and that the scholar ought to devote himself to his special researches without bothering to inquire whether they serve some purpose and lead anywhere.[83]

The absence of regulation may lead to pervasive personal disorganization. Beginning with Durkheim's study of suicide, most sociological interest in the concept of anomy has attached to the social deviance and personality disorganization it produces. Of the more recent sociological work in this area, Robert K. Merton's widely cited essay is still the most useful for the study of anomy in science.[84] His analysis is based on discrepancies between cultural goals (which can include such incentives as recognition) and institutionalized means for achieving them. He constructed five types of individual adaptations to such situations: (1) the conformist, who accepts both the goals and the means, (2) the innovator, who accepts the goals but rejects the use of legitimate means, (3) the ritualist, who rejects the goals but continues to accept the means, (4) the retreatist, who rejects both goals and means, and (5) the rebel, who rejects goals and means but substitutes new ones for them.

Durkheim's work on suicide and Merton's on deviation in the larger American society can be extended to the sciences. Although the evidence is somewhat incomplete, some material from interviews with mathematicians indicates that anomy is especially likely to be found in this discipline. The following discussion of anomy illustrates the theory and shows how the topic might be studied in greater detail.

THE DEVELOPMENT OF ANOMY IN MATHEMATICS

At first the mathematical disciplines were not sharply defined. As knowledge increased, individual subjects split off from the parent mass and became autonomous. Later, some were overtaken and reabsorbed in vaster generalizations of the mass from which they had sprung. Thus, trigonometry issued from surveying, astronomy, and geometry only to be absorbed, centuries later, in the analysis which had generalized geometry. This recurrent escape and recapture has inspired some to dream of a final, unified mathematics which shall embrace all. Early in the twentieth century it was believed by some for a time that the desired unification had been achieved in mathematical logic. But mathematics, too irrepressibly creative to be restrained by any formalism, escaped.[85]

A century ago, mathematics was divided into a number of relatively independent specialties—number theory, analysis, algebra, and geometry—and mathematicians could probably identify themselves and their colleagues easily. The twentieth century has seen a revolution in mathematics whereby the subject has become far more general and abstract, and the barriers separating the former specialties have been dissolved. As general theories transcending the old specialties have been constructed, the meaning of the specialties has become obscure.[86] But as barriers between specialties have been dissolved—as segments have been fused—specialization itself has become far more highly developed. It takes a mathematician a long time in a specialty to work to the "frontiers of research," and, once there, he seldom changes the direction of his work.[87] Thus, mathematics has come to the paradoxical situation of intense specialization without having clearly defined specialties. Many mathematicians find it difficult to identify their colleagues or themselves. The audience to which they address themselves is unknown or almost nonexistent. A mathematician has described the situation in language similar to that of Durkheim's description of the social science of his day: "Like the hordes and horses of some fabulous khan, today's mathematicians have ridden off in all directions at once, conquering faster than they can send messages home."[88]

EVIDENCE OF ANOMY IN CONTEMPORARY MATHEMATICS

Anomy in science can be specified as the general absence of opportunities to achieve recognition, and its accurate measurement would require interviews with large samples of scientists.[89] In the absence of such a survey of mathematicians, we must use indicators of anomy that are considerably less satisfactory. Some indicators are the perceived size of the typical audience, the frequency with which published papers are regarded as "trivial," the frequency with which mathematicians are concerned about being anticipated, and the difficulties experienced in evaluating the importance of specialties.

A number of mathematicians made similar statements; although they evidently involve some exaggeration, they were made without any apparent indication of facetiousness:

> A very large proportion of the things published are not read. The editor of *Mathematical Reviews* once said that he thought the average number of readers per mathematical paper was less than one.

A department chairman said:

> Mathematics is highly ramified; few people are in the same area. If you find three people reading your paper you feel flattered. . . . *It has been suggested that mathematicians form a tighter community than other scientists, a sort of guild.* . . . I doubt the business about community. Most mathematicians don't know of each other's work. You can use the figure of speech of the stream of science for physics, flowing down a relatively narrow channel. Mathematics is more like a delta. It is highly ramified. In mathematics you may have five people sharing neighboring offices, with neither able to understand the work of another. This department is more homogeneous, but I have a clear idea of the work of only three others. *But people attend the mathematics colloquia between X and Y universities.* . . . If you go to one, the number of mathematicians who understand the speaker is small. I'll grant that they get together to talk mathematics, but they don't understand one another's work.

A mathematician may have few colleagues, men to whom he can address his information. He may not be aware of any uses to which other mathematicians can put his work. A man working in lattice theory, a branch of abstract algebra, said:

> Research contacts tend to be rather few. I've had various members of the department come to me with various questions of an algebraic nature which tend to impinge on their work, but I've never had the feeling that it plays a vital role. You have the feeling that they think it would be nice to know it; you explain the area, and they understand it and go back to work. But I can't think of an instance when I've made a contribution that was of extreme significance to their work. *What about contributions in the other direction?* No . . . It is primarily because this work of mine is relatively isolated from other techniques. Techniques in analysis, for example, don't apply very well here. *Might techniques in topology?* No. There are certainly problems of this nature in topology, but I don't believe there is any transfer of techniques. If there were any, I think it would be from my techniques into those. But I don't think mine are powerful enough to do anything in the area of topology. . . . *Does it bother you that your work might not be appreciated by many people?* No, it doesn't. I don't know what the answer to that question is. I guess it doesn't bother me. . . . *Are you in contact with people doing similar work elsewhere?* A little bit. . . . There are probably three other people that I can think of offhand that I

> think are quite interested in this area. . . . [One of them] is a young fellow who has written some papers very similar to things I have done, and I have tried to get him interested to do more—I've written about some of the things I've been doing, and I can never get much of a rise from him, so I guess I can't count him.

Of course, there are other mathematicians who send reprints of their papers to hundreds of their colleagues, most of them personal acquaintances, and engage in discussions of mathematics with departmental colleagues and students most of the thirty or forty hours each week they spend around their university offices; these are leading men who work in "hot" or "fashionable" fields at leading universities.

But the mathematician as producer may often be unaware of any response by other mathematicians to his papers. In a complementary way, the mathematician as consumer may think that most of what he sees in the journals is "trivial." Perhaps the strongest statement of this type was made by an almost-applied mathematician, who said that many of his colleagues would disagree with him:

> [I] think the mathematician is spoiled by being left on his own too much, by being turned loose entirely on his own. . . . Because of sputnik so much money has been put into mathematics, and many people of low capacity have been turned loose. What they produce is often of no value, and others would agree with me in that. . . . It is very easy to do something new if one is not bound to do something specific. Much is done which is of little profundity or neatness. It is *new* and cannot be criticized on that ground. . . . In this country there is almost a catastrophe in mathematics. Pumping money into mathematics may have had a bad influence.

Even mathematicians in specialties with high prestige tended to agree. Two of them distinguished "trivial" from "uninteresting" work, agreeing that a great deal of the latter is published.

> There are two kinds of trivial research. First, there is the hard problem, involving the good use of techniques by competent mathematicians, but the work may be of little value. Second, there is the kind of problem which could be solved in two lines instead of five pages if the man stopped to think about it—and then he wouldn't publish it at all.

Such assertions may not be "objectively" correct. Whether or not published research is of potential significance for other mathematicians is not important here; what is important is that they often perceive it to be insignificant. When this happens frequently enough, a system in which information is exchanged for recognition is ineffective.

The definition of scientific anomy used here implies that a scientist in an anomic situation should not be concerned about being anticipated: if he does not expect others to find his problems important, he has no reason to expect them to work on the same problems. The fact that the mathematicians interviewed in this study are concerned about the possibility of being anticipated is evidence against viewing mathematics as a relatively anomic discipline. The number of responses is very small, yet suggestive: 64 per cent of fourteen mathematicians were concerned about being anticipated, as against 40 per cent of twenty-five physical scientists and 55 per cent of eleven molecular biologists.[90] Even some of the mathematicians who spoke about the low average readership of mathematical papers had been anticipated or were concerned about the possibility. Possibly mathematicians referred to their own experiences when discussing anticipation, but referred to other branches of mathematics when providing the statements on which the inference of anomy was based.

Another indication of anomy in a science is the absence of accepted criteria for ranking specialties according to importance. If specialties do not regularly contribute to each other, prestige differences between them are likely to be established on bases that will not be considered legitimate by many scientists. Relations between specialties in mathematics were more likely to be characterized by this dissensus than relations between specialties in other disciplines. In mathematics the disparaging assertion that a field is "fashionable" is frequently made, and mathematicians are more likely to complain about the low status awarded their specialty than scientists in other disciplines.

Many mathematicians would strongly object to the preceding characterization of their discipline, and they might attempt to explain it away by asserting that those of them who perceive a small audience for their work and cannot see the interesting aspects of the work of others are simply unqualified—doing uninteresting work themselves and not astute enough to be good judges of the work of others. This hypothesis need not be inconsistent with the theory presented here: mathematicians with less skill are more likely to be in anomic contexts. The assertion could still be made that there are more such persons in mathematics than in other sciences, and this is almost certainly not the result of a lack of talent among mathematicians.

More data would be necessary before we could make this assertion confidently. The sample of scientists interviewed for the present study was biased in the direction of eminence, so that scientists in anomic contexts were unlikely to be discovered. In addition, the interview guide was not constructed in a way to get material on "retreatism" with regard to scientific norms and goals. (To obtain such information from scientists would probably require depth interviews with actual deviants: most active scientists are unfamiliar with many instances of such deviance and unwilling to discuss them with strangers.) Even so, the material presented here is suggestive of the kinds of data that would be valuable in characterizing sciences as more or less anomic. Furthermore, the data lead to an explanation of certain types of behavior that seem to be found more often in mathematics than in other sciences.

CONSEQUENCES OF ANOMY IN MATHEMATICS

Mathematicians insist on the purity of their subject more than other scientists do:

> There is some kind of curious pride. A person will take care to tell you that he is not an applied mathematician. He may work on applied problems, but he's not *really* an applied mathematician.[91]

This stems partly from the low prestige awarded applied mathematics. But it is also the kind of behavior Durkheim suggested was typical in anomic societies, the insistence on the absolute value of personal goals and on the unrestrained freedom of the individual. This occurs in a different form in pure mathematics; while many mathematicians hope their work will interest colleagues in other specialties, others are indifferent. They assert that interest in mathematical topics is based not on the importance of results but on "aesthetic" factors: "In many ways mathematics is more of an art than a science. You might call it an exact art." The egoism of artists—also an anomic group in modern society—is paralleled by the egoism of mathematicians: "mathematics for mathematics sake" is sometimes the explicit slogan.

The assertion of the absolute value of one's personal goals does not mean the mathematician, any more than the artist, loses interest in social recognition. Rather, his need for recognition may grow as his estimate of the value of his own work grows, although it tends to be a covert need. The mathematician becomes more ambivalent with regard to recognition.[92] (The insistence upon the absolute value of individual goals may be a kind of defense reaction to this ambivalence.) Adaptations to a situation in which needs for recognition are continually frustrated may take the form of ritualism, retreatism, or rebellion.

The ritualist abandons the goal of receiving recognition for his contributions but continues to practice the institutionally prescribed means for achieving the goal, that is to say, he continues to contribute "information" in the usual way. In mathematics this means that much apparently trivial work will be published, as was noted above.[93] (Another explanation for the publication of trivial papers is that mathematicians must publish or lose their university positions. This explanation applies to persons having tenure only to a limited extent, and it fails to account for the apparent difference between mathematics and other sciences with regard to the amount of allegedly "trivial" work published. To the extent that mathematicians do publish for the sake of producing an impressive *vita* rather than to make their discoveries known to others, a substitution of goals occurs rather than ritualism or innovation; i.e., the behavior is akin to Merton's rebellion.)

Failure to be recognized, to have one's judgments reinforced by the judgments of others, may lead to a loss of faith in the value of one's work. When a man is highly specialized, and especially when he is old and dislikes the idea of spending years working into a new specialty, such loss of faith may be followed by a general withdrawal from creative work, a renunciation of both the goals and means of science. A number of mathematicians said they knew of such cases:

> This happens particularly among older mathematicians who started research along a particular line. Let me give you an example: At the turn of the century, a hot area in mathematics was the so-called theory of geometrical invariance. At the time, the notion of invariance was recognized as unifying a great many geometrical ideas. So there was an enormous study of invariance at the time. It was a fairly technical and fairly narrow area, and a lot of Ph.D.'s were being turned out at that time in the theory of invariance. Well, this went on for a time, and pretty soon they exhausted almost everything that could be said; the field wasn't closed, there were still a great many things to be done, but it was generally realized that what you were going to say would be more of the same kind of thing. It really wasn't going to bring entirely new ideas into the field. I've known a good many mathematicians, now all in their sixties, who were trained in this field, did their research in the area for many years, and didn't learn any of the related mathematics at the time. When the popularity of invariance theory went down they were simply left stranded. For example, they had difficulty getting papers published, because editors were just not interested in the material. So that they were perfectly aware that the field had passed them by, and they eventually gave up doing research. They felt they were too old to branch out into new fields and it was as plain as could be that it was impossible for them to receive recognition for research in what they had done. Obviously they were quite discouraged and frustrated.

Other mathematicians often advanced implicitly physiological arguments for the withdrawal of older men, asserting that mathematics is "a young man's game."[94] But there are too many instances of older mathematicians' making great discoveries or being otherwise productive for this to be convincing. The argument from physiology must be supplemented or even supplanted by the argument of anomic retreatism. Some idea of how the two fit together comes from the following statements from a young informant.[95]

> I'll stay in topology as long as I can do something. If I couldn't, I wouldn't stay in mathematics, I wouldn't want to. I might go into industry or another science. One's learning ability does drop off with age. This doesn't hurt in one's own field because although your ability drops your knowledge increases and you can still do very good work. . . . Short bursts of explosive

> energy are required [in mathematical research]. When one is old he is unable to do this—say after fifty. In my middle forties I might not feel happy staying in the subject. It may depend on how well I can do in developing students.

It is always difficult to work into a new area, and to do so requires both skill and motivation. Older mathematicians are less likely to change their special fields of interest. Perhaps this is because their abilities decline, although it seems equally plausible that their motivation declines: changing specialties means that they must postpone the gratifications of discovery for a considerable period, and during this period they must abandon the status they have achieved in a specialty and sit at the feet of other, perhaps younger, men who are experts in the new one. Having been disillusioned once, they may be unwilling to begin again to risk their status in competition with younger men.

Whatever the reasons for the disinclination of older mathematicians to change specialties, the effect may be the same: as the specialty declines in prestige, they find themselves in an anomic situation. Older men may adapt to anomy by retreatism, younger men by ritualism. Another possible reaction to anomy is rebellion.

The scientific rebel[96] rejects the goal of achieving recognition from members of the specialty in which he formally finds himself, and he rejects the means for doing so, namely, contributing information of a specific type; but, instead of withdrawing, the rebel substitutes a new community from which he desires recognition, and substitutes as means the contribution of a different type of information. This type of rebellion has already been discussed in this chapter.

Anomy and goal conflict therefore overlap, although they are not identical. In mathematics the conceptual revolution that broke down the existing system of specialization not only isolated individual specialists but also destroyed the existing relations between specialties. Ritualism and retreatism are individual responses to anomy, while specialty rebellion is a collective response, the response of a group whose relations with other groups are not regulated by the exchange of information.

Scientific specialties of insecure status may be especially anomic. If the specialty has no clear identity, its members will be dispersed, and the most frequent pathological adaptations will be ritualism or retreatism. As the specialty develops an identity and comes into conflict with the larger discipline, these adaptations may occur less frequently, to be replaced by rebellion. Reform specialties may thus be characterized by individual responses to anomy, rebellious specialties, by collective responses.

The remaining adaptation to anomy as described by Merton is innovation. In science this would take the form of commitment to the goal of recognition but attempting to attain it through illegitimate means. Perhaps the most common type of such innovation in contemporary science is the practice of publishing essentially the same discoveries in several different places. My impression is that this is less common in mathematics than in other disciplines. Other, more clearly deviant types of innovation are rare in mathematics. In part this is true for technical reasons. The mathematician, unlike the physicist or the sociologist, cannot trim or falsify his data or commit hoaxes;[97] his actual publication carries the evidence for his discovery in the form of a mathematical proof. Other types of deviation, such as plagiarism, are possible in mathematics but they are apparently just as rare in this discipline as in others.

CONCLUSION: ANOMY AND THE MARGINAL SCIENTIST

If a science is highly specialized and if specialties are independent of one another, recognition may not be awarded for the information contributed. This makes the scientist "independent" of his colleagues. This condition may be valued, with scientists attributing absolute value to their particular goals, although it may also be a source of frustration. Scientists will adapt to this frustration by ritualism, continuing to contribute information with no hope of its being recognized, retreatism, withdrawing from research, or rebelling against the research goals of the specialized community in which they find themselves.

A scientist's withdrawal from the community of research specialists usually means that he emphasizes other activities in which he can engage —especially teaching, but also administration of university and industrial laboratories, and perhaps sometimes leisure activities. These activities have rewards of their own that attract scientists in any case, and most discussions of withdrawal from research activity are in terms of conflicting role sets and career possibilities. Thus, it may be asserted that those who are unproductive of research give correspondingly greater emphasis to administration[98] or to teaching.[99] Explanations at this level are certainly important, but they imply that the reward system in science functions perfectly—that contributions by competent scientists are almost always rewarded. This explanation is insufficient; it must be complemented by a consideration of the reward system in science, one pathological aspect of which is anomy.

Anomy may have been more prevalent in early science than it is now. While the degree of specialization in early modern science was very low, the norms regarding recognition were correspondingly weak. Much the same may hold in contemporary nonscientific fields. In artistic and literary criticism, specialists are less likely to refer to one another's work than in science, and the control the colleague community can exercise over its members is correspondingly weaker.

NOTES

1. Cf. U. S. Bureau of Labor Statistics, *Occupational Mobility of Scientists*, Bulletin No. 1121 (Washington, D.C.: Government Printing Office, 1953). This study shows that a large proportion of scientists with the doctorate have worked in different disciplines. However, the study was not restricted to scientists in basic research but also included those in applied research, development, administration, and teaching. More interdisciplinary mobility must exist in applied research, and much of it must have resulted from occupational dislocations brought about by the depression and the mobilization of scientists in World War II. For a similar evaluation of this BLS study see David M. Blank and George J. Stigler, *The Demand and Supply of Scientific Personnel* (New York: National Bureau of Economic Research, 1957), pp. 105 f.
2. Cf. J. D. Bernal, *The Social Function of Science* (London: Routledge, 1939), pp. 262 ff., in which it is suggested that narrow specialization is more necessary for the biological sciences, where theory is weak, than for chemistry, where it is stronger. See also Diana M. Crane, "The Environment of Discovery" (Doctoral dissertation, Columbia University, 1964), pp. 83–89, 113–148.
3. This is also true in industrial research. One industrial research administrator encouraged his subordinates to work on more than one thing at a time, "since on one project alone they quickly reach the point of diminishing returns in the research." On the other hand, industrial scientists are often required to work on many problems for other reasons, and this is a frequent cause of tension in industrial research establishments. See William Kornhauser, *Scientists in Industry: Conflict and Accommodation* (Berkeley: University of California Press, 1962), pp. 68–70.
4. It is difficult for the layman to avoid error in discussing classifications. For example, Theodore Caplow and Reece McGee, in *The Academic Marketplace* (New York: Basic Books, 1958), pp. 92 f., write, "The evaluation of [the importance of a specialty] tends to be confused and emotional." To illustrate this, they cite two rather different evaluations by mathematicians of the importance of "topology." One mathematician was probably referring to "point-set topology," which is "old stuff," while the other was probably referring to "algebraic topology," a current fad.
5. Herbert Menzel, *The Flow of Information among Scientists* (New York: Columbia University Bureau of Applied Social Research, 1958), p. 35. Diana Crane asked her sample of biologists, psychologists, and political scientists how many other people in their fields worked on "types of problems" similar to their own; 87 per cent of her respondents could make an estimate, most said such groups consisted of fewer than twenty-five members, and many said the groups included fewer than six. Crane, *op. cit.*, p. 175.
6. See Derek J. de Solla Price, *Little Science, Big Science* (New York: Columbia University Press, 1963), ch. I. Price points out that more than 80 per cent of the scientists who have ever lived are living now.
7. Cf. *ibid.*, pp. 83–86.

8. Charles C. Davis, "Biology Is Not a Totem Pole," *Science*, 141 (1963), 308–310.
9. Cf. Richard L. Meier, "The Origins of the Scientific Species," *Bulletin of the Atomic Scientists*, 7 (1951), 172.
10. Cf. Barry Commoner, "In Defense of Biology," *Science*, 133 (1961), 1745–1748.
11. These criteria are taken from Herbert Feigl, "The Scientific Outlook: Naturalism *and* Humanism," in Herbert Feigl and May Brodbeck, eds., *Readings in the Philosophy of Science* (New York: Appleton-Century-Crofts, 1953), pp. 8–18.
12. The hypothesis that the prestige of a field is highly correlated with the mean IQ of recipients of the doctorate within it is not supported by data, however. E.g., one study showed that Ph.D.'s in engineering and the social sciences had a higher mean IQ than Ph.D.'s in chemistry or biology. See Lindsey R. Harmon, "The High School Backgrounds of Science Doctorates," in National Science Foundation, *Scientific Manpower 1960* (NSF 61–34, Washington, D.C.: Government Printing Office, 1961), pp. 14–26.
13. Meier, *op. cit.*, p. 172.
14. The interview took place shortly before Mössbauer received the 1961 prize for his work in low-energy physics.
15. See Thomas S. Kuhn, *The Structure of Scientific Revolutions* (Chicago: University of Chicago Press, 1962), chs. III and VI; and Gerald Holton, "Scientific Research and Scholarship: Notes toward the Design of Proper Scales," *Daedalus*, 91 (1962), 362–399.
16. Or, as Thorstein Veblen would put it, invidious distinction and emulation. See *The Theory of the Leisure Class* (New York: New American Library, 1953), pp. 81 f. *et passim*. See also the general treatment of fashion in Georg Simmel, "Fashion," *American Journal of Sociology*, 62 (1957), 541–558.
17. A. L. Kroeber, *Anthropology* (rev. ed.; New York: Harcourt Brace, 1948), pp. 329–336.
18. Honor B. Fell, "Fashion in Cell Biology," *Science*, 132 (1960), 1625–1627.
19. Cf. the discussion in Robert K. Merton, "Resistance to the Systematic Study of Multiple Discoveries in Science," *European Journal of Sociology*, 4 (1963), 237–282, especially pp. 267 ff.
20. The size of the community also makes a difference. As Veblen pointed out, conspicuous consumption is unlikely to be important in small communities, but when the community is larger and "the means of communication and the mobility of the population now expose the individual to the observation of many persons who have no other means of judging of his reputability than the display of goods," conspicuous consumption becomes more important. Veblen, *op. cit.*, p. 71. In other words, fashion in general becomes important only when what Durkheim called the "moral density" of communities becomes great. Since most scientific communities are now quite large, their relative sizes cannot account for differences in fashion among them.
21. This can be measured, at least crudely, by compiling the distribution of dates of references appearing in research reports in the various sciences. Price refers to studies that show that half of all references in chemistry were made to papers less than eight years old, half in physics to papers published in the preceding five years, and half in *Physical Review Letters* to papers appearing in the preceding two and one half years. D. J. de Solla Price, *op. cit.*, pp. 78–82.
22. Cf. a statement by the mathematicians in the Progressive Education Association: "In reflecting the spirit of the times, mathematics follows fashions. . . ." In

William L. Schaaf, ed., *Mathematics: Our Great Heritage* (New York: Harper, 1948), p. 238. Other mathematicians have expressed the same idea; see Eric T. Bell, *The Development of Mathematics* (2nd ed.; New York: McGraw-Hill, 1945), pp. 206–211 *et passim;* and Norbert Wiener, *Ex-Prodigy: My Childhood and Youth* (New York: Simon and Schuster, 1953), p. 230.

23. Some exceptions to this occur when scientists are completely wrong. For example, stimulated by the discovery of X rays, a French physicist discovered "N rays" in 1903. "Once the start had been made the new field grew rapidly. Nearly one hundred papers on N rays were published in the official French journal *Comptes Rendues* during 1904. . . . A success so resounding could not go unrewarded, and eventually, in the same year, the French Academy decided to honor their new discoverer with the considerable Leconte Prize of 20,000 francs and a gold medal." Within the year an American expert "showed it reasonable to attribute all the subjective effects to wishful thinking and to the overpowering difficulty of estimating by eye the brightness of faint objects. . . . From that day onward, there were no N rays." Derek J. de Solla Price, *Science since Babylon* (New Haven: Yale University Press, 1961), pp. 85 f., 88.

24. I do not know of hard data with which to substantiate this assertion. See the informed observations reported by G. H. Hardy, *A Mathematician's Apology* (Cambridge: Cambridge University Press, 1941), p. 10; and C. Eisenhart in E. S. Pearson, "Report of the Committee on the Supply and Demand for Statisticians," *Journal of the Royal Statistical Society*, Series A, 122 (1959), 73.

25. Cf. Albert Einstein's statement in Paul A. Schilpp, ed., *Albert Einstein: Philosopher-Scientist* (Evanston, Ill.: Library of Living Philosophers, 1949), pp. 15 f.

26. *Ibid.*, pp. 87 f. See also the comments by Wolfgang Pauli, Max Born, and Niels Bohr on Einstein's intransigence, in the same volume. Born wrote, "Many of us regard this as a tragedy—for him, as he gropes his way in loneliness, and for us who miss our leader and standard-bearer. I shall not try to suggest a resolution of this discord. We have to accept the fact that even in physics fundamental convictions are prior to reasoning, as in all other human activities." *Ibid.*, pp. 163 f.

27. See the sketch of Cantor's tragic life in Eric T. Bell, *Men of Mathematics* (New York: Simon and Schuster, 1937), ch. 29.

28. "Harold Hotelling—A Leader in Mathematical Statistics," in Ingram Olkin *et al.*, *Contributions to Probability and Statistics: Essays in Honor of Harold Hotelling* (Stanford: Stanford University Press, 1960), p. 9.

29. Commoner, *op. cit.*, p. 1747.

30. *Ibid.*, p. 1746.

31. "Dethronement," *The Magazine of Fantasy and Science Fiction*, 21 (November 1961), 71. (This article was written as a rebuttal to Commoner's paper.)

32. Commoner, *op. cit.*, p. 1745.

33. A "collegial" organization is one in which decisions are made by a group of persons of more or less equal status who have secure claims to their positions; it can be contrasted with "monocratic" authority ("one man rule") and with bureaucratic forms of organization. See the discussion in Max Weber, *The Theory of Social and Economic Organization*, A. M. Henderson and T. Parsons, trans. and ed. (Glencoe, Ill.: Free Press, 1947), pp. 392–404.

34. Compare what David Riesman wrote about disciplines as veto-groups in the larger university structure: *Constraint and Variety in American Education* (Garden City, N.Y.: Doubleday Anchor, 1958), pp. 107–114.

35. See, for example, William V. Consolazio, "Dilemma of Academic Biology in Europe," *Science,* 133 (1961), 1892–1896; R. P. Grant, C. P. Huttrer, and C. G. Metzner, "Biomedical Science in Europe," *Science* 146 (1964), 493–501.
36. (Garden City, N.Y.: Doubleday Anchor, 1959).
37. Acquiring such autonomy is a considerable victory for the deviant specialty. The decision to grant autonomy is not made by the department alone but also involves other departments and administrators in the university, since it is alleged that the deviant specialty is the proper interest of more than one department. In other words, when the question of establishing something like a "group curriculum" or "area major" is raised, it probably manifests the efforts of the deviant specialty to escape the constrictions of its discipline by appealing to higher authorities.

    Giving the major advisor almost complete authority over the dissertations done under his supervision is one way of making the departmental system flexible and circumventing the control of "traditionalists." It is thus similar in function to the "elective system," in which autonomy was given to students in an effort to break the stranglehold of the classics on the undergraduate curriculum. This is apparently what happened at Harvard, for example; see G. W. Pierson, "American Universities in the Nineteenth Century," in Margaret Clapp, ed., *The Modern University* (Ithaca, N.Y.: Cornell University Press, 1950), pp. 59–94, especially pp. 85 f.
38. *The Logic of Liberty* (London: Routledge and Kegan Paul, 1951), pp. 53–55.
39. Richard Colvard, "Foundations and Professions: the Organizational Defense of Autonomy," *Administrative Science Quarterly,* 6 (1961), 167 f.
40. Universities lack this flexibility—i.e., universities must be more strongly committed to existing organizational forms. A foundation representative suggested that university officials might prefer not to dispense research funds themselves because of this lack of flexibility and because it "relieved them of the difficulty of drawing invidious distinctions within their own faculties." Quoted in Dael Wolfle, ed., *Symposium on Basic Research* (Washington, D.C.: American Association for the Advancement of Science, Publication no. 56, 1959), p. 236.
41. In practice the research organizations of the Soviet Union appear to be flexible. This is especially evident in the area of scientific education; see Alexander G. Korol, *Soviet Education for Science and Technology* (New York: Wiley, 1957), pp. 155 f., 215, *et passim.* Although there have been complaints that barriers between the different academies of the Soviet Academy of Science inhibit interdisciplinary research, and although Soviet science has been undergoing extensive reorganization, this apparently has resulted from the general problem of the relations between basic and applied research, not the specific organization of basic research; see Nicholas De Witt, "Soviet Science: the Institutional Debate," *Bulletin of the Atomic Scientists,* 16 (1960), 208–211, and "Reorganization of Science and Research in the U.S.S.R.," *Science,* 133 (1961), 1981–1991. If disciplinary organization in the Soviet Union is highly flexible, it probably results from the Soviet disposition to see science as an instrument for achieving other ends, rather than as an end in itself.
42. Robert K. Merton, *Social Theory and Social Structure* (Glencoe, Ill.: Free Press, 1949), p. 145.
43. E.g., Paul E. Klopsteg, "The Indispensable Tools of Science," *Science,* 132 (1960), 1913–1922; Joseph H. Simons, "Scientific Research in the University," *American Scientist,* 48 (1960), 80–90; Bernal, *op. cit.,* p. 78, *et passim;* and, of course, Auguste Comte, *The Positive Philosophy,* freely trans. and condensed by

Harriet Martineau (vol. 2; London: John Chapman, 1853), pp. 459 f. *et passim.* About half the department chairmen interviewed were rather apologetic about their positions, stressing that their productivity was far less than it had been before they took their positions and making such statements as, "I am an inefficient administrator," or "I'm essentially the executive agent of the department —This is awful occasionally." Administrative skills are not highly valued in the scientific community.

44. Cf. Chester I. Bernard, "Functions of Status Systems in Formal Organizations," in Robert Dubin, ed., *Human Relations in Administration* (Englewood Cliffs, N.J.: Prentice-Hall, 1951), pp. 264–267; Robert W. Lamson, "The Present Strains between Science and Government," *Social Forces,* 33 (1955), 360–367.

45. Scientists who express themselves primarily in books can be expected to have distinctly different orientations with respect to competition, their reference groups, and their modes of conducting disputes. A statistician interviewed was unusually forceful about putting controversial propositions in print. He asserted, "Nothing is more powerful than a book to put the stamp of authority on a theory. . . . With a book there are no limitations of time or space. A book will either convince or not convince. . . . A book gets wider circulation. Papers only get into the hands of the cognoscenti. Books get into the hands of those in other fields."

46. There were almost 100,000 scientific periodicals at the beginning of the fifties, including almost 300 journals devoted solely to abstracting scientific publications. See the references in D. J. de S. Price, *Little Science, Big Science, op. cit.,* pp. 95–101. In 1950, there were 151 journals in physiology alone; see K. E. Rothschuh and A. Schafer, "Quantitative Untersuchungen über die Entwicklung des Physiologischen Fachschrifttums (Periodica) in den Letzen 150 Jahren," *Centuarus,* 4 (1955), 63–66.

47. Lancelot Hogben, *Statistical Theory* (London: George Allen and Unwin, 1957), pp. 235 f.

48. See Helen M. Walker, "The Contributions of Karl Pearson," *Journal of the American Statistical Association,* 53 (1958), 18; and E. S. Pearson, "Karl Pearson: Some Aspects of His Life and Work," *Biometrika,* 28 (1936), 231 f.

49. *The Life and Work of Sigmund Freud* (New York: Basic Books, 1957), III, pp. 31–38. See also Jones on the importance of its first journal to psychoanalysis in 1910; *op. cit.,* (1955) II, p. 45.

50. R. W. Gerard suggests that the innovators of new disciplines tend to be young, highly committed, even zealots, since otherwise the conservatism of the old cannot be overcome. "Problems in the Institutionalization of Higher Education," *Behavioral Science,* 2 (1957), 134 f. See also David Riesman, *op. cit.,* pp. 105–107; there he suggests that converts to a discipline may be more serious about disciplinary commitments than others. He cites a study of graduate students in physiology on this point: Howard S. Becker and James Carper, "The Elements of Identification with an Occupation," *American Sociological Review,* 21 (1956), 341–348. Riesman also suggests that the fervor of this disciplinary nationalism "reflects the sacrifices the scholar has made to become a scholar, what he has had to surrender of earlier social-class origins and ambitions." In France and England there is apparently less identification with disciplines because of early academic specialization, and a tendency among students to identify more with particular professors than with disciplines.

51. Cf. Kuhn, *op. cit.,* ch. 11.

52. To say that such historical writings are "mythical" is not to imply that they necessarily err in the presentation of facts, although this is common; see *loc. cit.* Rather, their function is to help scientists identify themselves and to reinforce their commitments to scientific values, and information is selected and interpretation distorted to achieve this. On this idea of myth see Robert M. MacIver, *The Web of Government* (New York: Macmillan, 1947), pp. 4, 39, *et passim.*

53. Sociologists will be familiar with the claims of Auguste Comte in this regard; see Comte, *op. cit.*, pp. 485 ff., 538 f.

54. Paul A. Samuelson, "Harold Hotelling as Mathematical Economist," *American Statistician*, 14 (1960), 25.

55. Sir Ronald Fisher's text, *Statistical Methods for Research Workers*, which went through thirteen editions between 1925 and 1958, is perhaps one of this century's most influential books.

56. J. G. Crowther, *British Scientists of the Nineteenth Century* (London: Kegan Paul, 1935), pp. 4 f.

57. Cf. George Herbert Mead's treatment of the relation between Romanticism and the development of national self-consciousness in nineteenth-century Europe, *Movements of Thought in the Nineteenth Century* (Chicago: University of Chicago Press, 1936).

58. Hans Freudenthal, "The Main Trends in the Foundations of Geometry in the 19th Century," in Ernest Nagel, Patrick Suppes, and Alfred Tarski, eds., *Logic, Methodology and Philosophy of Science* (Stanford: Stanford University Press, 1962), pp. 613–621.

59. Alfred North Whitehead, quoted in Robert K. Merton, *op. cit.*, p. 3. It is worth noting that Whitehead himself wrote an influential popular history of modern science (with a moral, of course): *Science and the Modern World* (New York: Macmillan, 1925).

60. This distinctiveness is often exaggerated by the formation of a jargon that is probably not technically required. Karl Pearson, for example, was fond of coining Greek words to describe statistical distributions. The jargon of sociologists and psychoanalysts has often been noted by nonspecialists. Jargon also has the function of isolating the emerging discipline from a hostile environment.

61. Joseph Ben-David, "Roles and Innovations in Medicine," *American Journal of Sociology*, 65 (1960), 557–568.

62. On biochemistry, see R. H. A. Plimmer, *The History of the Biochemical Society 1911–1949* (Cambridge: Cambridge University Press, 1949). Biophysics has already been mentioned.

63. A. W. Coats, "The First Two Decades of the American Economic Association," *American Economic Review*, 50 (1960), 555–574.

64. "Statisticians—Today and Tomorrow," *Journal of the American Statistical Association*, 54 (1959), 1–11.

65. Freud is the clearest example of this. Most leading psychoanalysts of his time had been analyzed by him; he had been their teacher and the discoverer of the theory they accepted; he aided them materially by referring patients to them; and he played an active role in their congresses and publications.

66. Freud's strained relations with many of his followers have often been noted. Even the greatest admirers of Karl Pearson admit his authoritarian proclivities; see, e.g., E. S. Pearson, "Karl Pearson: Some Aspects of His Life and Work," *Biometrika*, 29 (1938), 184.

67. "Organized Research Units," in *University Bulletin* (California), 10 (1962), 112–113.
68. Bernard Berelson, *Graduate Education in the United States* (New York: McGraw-Hill, 1960), p. 35.
69. See the discussion of German universities in the nineteenth century in Joseph Ben-David and Awraham Zloczower, "Universities and Academic Systems in Modern Societies," *European Journal of Sociology*, 3 (1962), 54–56. See also Norman Kaplan, "The Western European Scientific Establishment in Transition," *American Behavioral Scientist*, 6 (December 1962), 17–21.
70. See Kornhauser, *op. cit.*, pp. 50–56, for a more detailed discussion and references to the literature.
71. See *ibid.*, pp. 149–154.
72. The ideas about social change presented here have been informed by Neil Smelser's treatment of the topic: *Social Change in the Industrial Revolution* (Chicago: University of Chicago Press, 1959). Because of the nature and relative paucity of the data collected, his approach could not be followed closely.
73. Cf. Émile Durkheim's concept of moral density, presented in *The Division of Labor in Society*, George Simpson, trans. (Glencoe, Ill.: Free Press, 1947), pp. 256–264.
74. Charles Darwin, *The Origin of Species* (6th ed., New York: Mentor, 1958), p. 82.
75. Garrett Hardin, "The Competitive Exclusion Principle," *Science*, 131 (1960), 1292–1297.
76. Darwin, *op. cit.*, p. 106.
77. This scheme would seem to be applicable to other types of segmentation in communities, such as the segmentation of religious groups, families, and nations.
78. "The Tree of Knowledge," *Harper's*, 217 (October 1958), 55, 57. (Some italics deleted.) See also Oppenheimer's *Science and the Common Understanding* (New York: Simon and Schuster, 1954).
79. *Division of Labor, op. cit.*, p. 368.
80. *Ibid.*, p. 370.
81. *Ibid.*, pp. 369 f., and *Suicide*, John A. Spaulding and George Simpson, trans. (Glencoe, Ill.: Free Press, 1951), pp. 241–276.
82. Durkheim, *Division of Labor, op. cit.*, p. 357.
83. *Loc. cit.*
84. *Social Theory and Social Structure* (Glencoe, Ill.: Free Press, 1949), ch. IV.
85. Eric T. Bell, *The Development of Mathematics, op. cit.*, p. 20. Bell suggests that group theory played a role similar to mathematical logic in the latter part of the nineteenth century.
86. See *ibid.*, p. 326, on the problem of defining geometry today.
87. Cf. Robert D. Carmichael, "The Larger Human Worth of Mathematics," in Schaaf, *op. cit.*, p. 271.
88. David Bergamini, quoted in a book review by Wallace Givens, *Science*, 142 (1963), 1287.
89. Even with such a survey it would be difficult to measure anomy. It would be necessary to obtain estimates from scientists of their confidence in obtaining recognition from the research alternatives open to them. Anomy would be a characteristic of the distribution of such estimates for the scientists in a discipline or part of a discipline; while the mean degree of confidence is the obvious way

of summarizing such a distribution, it is not the only way. It is clear that different parts of a discipline could be more or less anomic. Durkheim noted that this was typically true for the larger community; see, e.g., *Suicide, op. cit.*, p. 245.

90. See also Chapter II, p. 75. There, for the reasons indicated, mathematicians were grouped together with other formal scientists.
91. Some mathematicians complain about their colleagues' exaggerated stress on purity; see, e.g., Norbert Wiener, "Science and Society," *Technology Review*, 63 (July 1961), 49–52.
92. Scientists are generally ambivalent with regard to seeking recognition. See Chapter I, and Robert K. Merton, "Resistance to the Systematic Study of Multiple Discoveries in Science," *op. cit.*, pp. 267 ff.
93. The argument here may appear to be circular. The same datum, the alleged presence of trivial research, which has been used here as evidence of a typical consequence of anomy, was used above as evidence for the existence of anomy. Since, however, there are other reasons for characterizing mathematics as anomic, not all of which are then "explained" by anomy, it is legitimate to use these data at two points in the argument.
94. G. H. Hardy, *op. cit.*, p. 10. The available data do not support Hardy's assertion; there is no good evidence to show that productivity declines with age in mathematics or any other field of science. Harvey C. Lehman presented some data which seemed to suggest this in his *Age and Achievement* (Princeton, N.J.: American Philosophical Society, 1953). However, his findings have convincingly been attacked as statistical artifacts. See Wayne Dennis, "The Age Decrement in Outstanding Scientific Contributions: Fact or Artifact?", *American Psychologist*, 13 (1958), 457–460. On the basis of data he collected, Dennis concluded, "productivity persists in the later years of life;" "Age and Productivity among Scientists," *Science*, 123 (1956), 724–725. Similar results are reported in C. W. Adams, "The Age at Which Scientists Do Their Best Work," *Isis*, 36 (1945), 166–169.
95. See also Norbert Wiener, *I Am a Mathematician* (Garden City, N.Y.: Doubleday, 1956), pp. 42 f.
96. "Rebel" in the sense used by Merton.
97. See the discussion of these types of deviation in Chapter II.
98. Cf. Simon Marcson, *The Scientist in American Industry* (New York: Harper, 1960), pp. 69 f. In one highly regarded research organization, the tendency of supervisors to be less committed to research existed only among those of them who received relatively little recognition for their research accomplishments. See Barney G. Glaser, *Organizational Scientists: Their Professional Careers* (Indianapolis, Ind.: Bobbs-Merrill, 1964), ch. 5.
99. Cf. Alvin W. Gouldner, "Cosmopolitans and Locals: Toward an Analysis of Latent Social Roles," *Administrative Science Quarterly*, 2 (1957), 293–296.

V

STRUCTURAL CHANGE: FUNCTIONAL DIFFERENTIATION

Disciplinary differentiation is a kind of segmentation. The "segments" formed have similar internal structure, but differing goals. This makes it possible for the differentiated units to be socially autonomous. The units are not dependent on one another in the short run; scientists seldom need to pay close attention to research done outside their disciplines, and they rarely need to seek assistance from members of other disciplines.

Informants were asked many questions about their relations with scientists in other disciplines, and their responses indicated that cross-disciplinary consultation is rare. Physicists, for example, seldom consult mathematicians; when they do, it is usually because they have mathematicians among their personal friends, and the advice they receive is usually limited to a reference to the literature. Similarly, physical chemists and metallurgists, unless they happen to be "chemical physicists," seldom consult solid-state physicists.

Of course, there is a sense in which science forms an interdependent unity; discoveries made in one discipline may have ramifications throughout all science. But—with significant exceptions—the essential knowledge a scientist needs of other disciplines can be acquired while he is a student, even while he is an undergraduate student. It can be argued, of course, that disciplines are too isolated from one another, that progress is delayed because scientists are not aware of discoveries made in other disciplines which would have bearing on their own research. This was one of the major arguments of Auguste Comte.[1] Recent history leads me to be skeptical of these claims. For example, Linus Pauling and others applied quantum mechanics (physics) with great success to chemistry by the early nineteen thirties, although the physical theories were only formulated in the period 1925–1927 by Schrödinger, Heisenberg, and others. In the past decade, X-ray diffraction techniques have been applied to genetic material, indicating rapid diffusion of procedures devised by physicists and physical chemists in the years following 1912. Many other examples could be cited. In any case, for my purposes it is enough that scientists themselves define the situation as one where disciplines have little short-run interdependence.

Functionally differentiated elements are interdependent in the short run. Discovery in one element has immediate or almost immediate effects on work in the other. When research in various branches is so closely related, it may be felt proper for even graduate students to have the same course of instruction, regardless of their specialties. This prevents segmentation, but the discipline may become functionally differentiated. The differentiated elements form parts of the same organizational units: roles are differentiated; research units seldom are.[2]

Functionally differentiated elements are rare in science,[3] and the only really well-developed example is the differentiation of physics into "theoretical" and "experimental" components. There are two conditions for such differentiation. The first is the logical elaboration of abstract theory. This makes it necessary to receive special training in order to deal creatively with theories; it also means that theoretical work usually requires the full-time efforts of individuals ("necessary" and "requires" in the sense that those who do so specialize will be able to accomplish more in a competitive situation than those who do not). The second condition is the growth of technological complexity in the collection of data by experiments and field research. Special training and continual study is necessary, in the same way, in order to cope with this complexity.

If data-collection becomes highly complex but theory does not, differentiation will not occur. Experimenters will resist the formation of a differentiated group of theorists, although a manager interested in the rationalization of labor (such as Francis Bacon in *The New Atlantis*) might believe that differentiation would increase efficiency. Experimenters think this way because functionally differentiated theoretical roles typically acquire higher prestige. Furthermore, experimenters will attempt to "complete" their research and will resent the utilization of it by theorists, who "haven't done the work." Data from interviews with physical chemists, organic chemists, and molecular biologists illustrate this resistance.

There are a few theoretical physical chemists who do essentially what theoretical physicists do, only on chemical systems. One of them, on being asked whether there was any resistance to such specialization, said:

> There used to be, I understand, before the war, but the resistance has broken down now. . . . There may still be some hidebound departments, but none that I know of. . . .

Another physical chemist, who did primarily experimental work, said of theoretical physical chemists (erroneously, I believe):

> It is not true to say that they do no experiments at all. All get involved in experiments: they try to talk someone else into doing the experiment for them, and, if they can't, they do it themselves. One of the few theoretical papers I wrote was when I was in that position—I didn't want to do the experiment.

Organic chemists show more resistance to purely theoretical roles. One of them said:

> There are very few purely theoretical papers; it is hard to get them published. The feeling in the field seems to be that anybody can have ideas. Perhaps one could publish a paper based on correlations of previously published experimental results; even this isn't very common, however.

Some of the journals in experimental biology make their resistance to theoretical or, as the specialists concerned call it, "speculative," work quite explicit. Thus, one leading journal states in its "Notice to Authors": "Conclusions should be based on the experimental data submitted. . . . The experimental procedure must be clearly described."[4] Another journal is more permissive: "Articles with speculative contents are appropriate for publication only if they will stimulate research from a new viewpoint."[5] And recently at least one (inexpensively reproduced) journal has been established to permit rapid publication of both experimental and theoretical work.[6] However, as one informant said, though "some minor journals" do accept theoretical papers, "these are not looked upon especially highly."

Molecular biologists were asked whether a specialist could base a good reputation solely on theoretical work, and most of them asserted without qualification that this was impossible. They were scornful of the efforts of physicists who have attempted to do purely theoretical work in the area:

> When the physicists first moved into biology, twenty or thirty years ago, they often made a purely theoretical approach. This has been useful essentially to stimulate other physicists to become interested in it. Actually, it contributed nothing at all to real progress in the field. But some physicists have become essentially experimental biologists.

Some molecular biologists have, however, acquired fame for theoretical work alone. Linus Pauling won the Nobel Prize in chemistry (1954) partly because of his theoretical contributions to protein chemistry. F. H. C. Crick won a Nobel Prize in 1962 for his theoretical contributions to the problem of coding genetic information in molecules of DNA (deoxyribonucleic acid).[7] The response of an informant confronted with this exception illustrates the nature of the resistance to theorizing in molecular biology:

> *Do you think it is possible to get a reputation for good work in molecular biology for theoretical work without experimental work?* I don't see how. *What about Crick?* I don't think that's fair to Crick. I know the details of the work that got him his reputation. He worked hard, and it was not just inspiration. It doesn't matter if you call it experimentation. It took a couple of years of very intensive model-building, a sort of unique kind of experimental work. He didn't collect the data himself, but worked just as hard as if he had.

A scientist's experimental discoveries are his property, and it is felt to be a kind of theft to utilize the experimental data of others for purely theoretical work. There are, of course, more sophisticated arguments against differentiating a theoretical role; one informant spoke on the point at some length and concluded by saying: "I think it is generally recognized that it is almost impossible to evaluate the significance of experiments in a field where one is not very active himself, because it is necessary to know how it is done, what are its drawbacks, how it could be wrong, and so on." The same argument could be made in physics; but, despite this, the roles of theorist and experimenter are clearly differentiated.

Though role differentiation will not occur in the absence of highly developed mathematical theories, neither will it in the absence of a complex data-gathering technology, for then the data needed by theorists will be available to everyone or can be obtained by the theorist himself or his relatively untrained assistants.[8]

Physics

In contemporary physics, the role of the theorist is almost completely differentiated from that of the experimenter. (The differentiation is most complete in nuclear physics and least complete in such areas as solid-state physics. In the latter, experimenters are often able to be theorists as well.) Theorists are concerned with the specification and extension of existing theories to new empirical areas and with developing new theories, whereas experimenters are concerned with testing theories and with obtaining data in new areas. There are many more experimenters than theorists. Specialization occurs in the course of graduate training; although students may sometimes choose one or another role earlier, they must decide by the time they come to prepare a dissertation. Few individuals change from one role to the other in the course of their careers. By and large, the college curriculum for theorists and experimenters is the same; a few courses may be taught only for theorists, and preliminary examinations for the Ph.D. differ. The content of most academic courses is theoretical, and the skills of the experimental physicist are acquired in an apprenticeship situation. Either experimenters or theorists may teach most courses offered, although some subjects may be taught mostly by theorists. Training, then, is undifferentiated in all but the most advanced stages, whereas research work is almost completely differentiated.

Theorists and experimenters belong to the same scientific societies, and the major journals are also undifferentiated. (It is difficult for a layman to identify a physicist as a theorist or an experimenter; they describe their "specialties" in such directories as *American Men of Science* with the same terms, they may teach the same undergraduate courses, and the titles of the papers they write are in both cases difficult to understand.) Within such research establishments as nuclear-physics laboratories, there are usually separate groups of theoretical physicists, however.

This differentiation has been evolving for more than a century. In the eighteenth and early nineteenth centuries, the roles of physicists and mathematicians often overlapped. In the course of the nineteenth century, experimental technology became more complex and eventually was included in university instruction. Though brilliant experimental scientists like Michael Faraday and J. P. Joule did their work outside universities, the work of each was closely linked with mathematically trained physicists who were academics. Joule collaborated with William Thomson (Lord Kelvin), and Faraday's experiments and intuitive schemes were eventually cast into a mathematical theory by James Clerk Maxwell. But Max Planck's experiences in the eighteen eighties indicate that, at that time, the role of the pure theorist was not well established in European universities.

> While an instructor in Munich, I waited for years in vain for an appointment to a professorship. Of course, my prospects for getting one were slight, for theoretical physics had not as yet come to be recognized as a special discipline.[9]

A generation later, the differentiation appeared "traditional" to Leopold Infeld. In the nineteen twenties, he sought a theoretical post in Poland; writing of it, he mentions: "It was a tradition on the Continent that each university should have at least two professorships in physics; one in theoretical and one in experimental physics."[10] It is beyond the scope of this work to treat the history of the differentiation of these roles in detail. Further examination of historical data would be valuable, first, to describe the resistance by universities to the incorporation of experimental physicists, men who worked "with their hands," and, second, to describe the resistance by experimental physicists to the differentiation of a theoretical role.

There was certainly resistance by experimenters to role differentiation. This has been portrayed vividly by Mitchell Wilson in his novel, *Live with Lightning*.[11] It depicts an experimental physicist who is extremely hostile to theorists and uses his power against them.[12] Even now there are strains between theorists and experimenters. At one university (not a leading institution), a theoretical physicist accounted for the relative youth of the theorists in his department by saying:

> You have hit a sore point there. The senior members are all experimental physicists—all the associate and full professors are. Since they make all decisions on policy and appointments, the situation is one of insecurity for theoretical physicists. As a result, the turnover of theorists here is very large, and this will continue until the balance is changed.

These strains result from prestige differences, problems of communication, and the proprietary interests of experimenters in the data they obtain. A former physicist referred to theorists and experimenters as the "elite" and the "peons," respectively.[13] (However, experimenters are usually more powerful in formal organizations. For example, directors of physics laboratories are more likely to be experimenters than theorists. This is so because there are more of the former and because they are more accustomed by the nature of their research to "administrative" and "political" work.) The difference in status is sometimes expressed but usually denied by experimental physicists. Thus, a graduate student chose to be an experimental physicist after his first year of graduate school because:

> I didn't know too much about theory or experiment, and my advisors said I'd be happier as an experimenter. I thought then one had to be smarter to be a theoretician; I learned later that this isn't so—it helps being smart to be either an experimenter or a theorist.

Fame outside the discipline is most likely to be achieved by theorists, who have constituted most of the recent popular heroes of physics.

Some experimental physicists are unhappy about their inability to do theoretical work, not so much because of the status difference as because of its effects on the conditions of their work. Enrico Fermi may possibly be a folk-hero to them because he successfully combined the two roles.[14] Emphasis on careers like his tends to obscure the extent of functional differentiation, and some experimenters may aspire to do likewise. A student liked the area in which he was doing his dissertation research because the theory for it was relatively undeveloped:

> If a theory is available, it directs you to just a few results. It's harder to start from the physical observables and try to understand them. . . . Given a theory of nuclear interactions, you can often find that it explains many par-

ticular interactions. After a while, you get sick of experiments on it, since there is less thought involved. . . . It seems to me that 80 per cent of such experiments is just *doing*—getting the experiment to work—and 20 per cent is interpretation, but interpretation along known lines. You know where you're going. On the other hand, when no theory exists, you must explore all possibilities. . . . *Is it clear to you in your work where the experimental physicist ends his work and the theorist begins?* Yes, it's clear, but I don't like it. I would like it if all physicists were both.

Theoretical physicists may be scornful of such attempts:

To a large extent I think [the interpretation of experiments] should also be left to the theorists. Experimentalists are apt to bungle results and present them in a form which is incomprehensible to the theorist—not only incomprehensible; sometimes they simply remove most of the information.

But theorists themselves may resist being called on to interpret an experiment. Thus, in one nuclear-physics laboratory, "There is some service work [by theorists]; some do the dirty work of computations. X, head of the theoretical group, does make assignments of that sort."

Theorists themselves are differentiated. Some—sometimes called "phenomenological theorists"—work directly with experimental data; some—called "mathematical physicists" or "formalists"—work at very abstract levels; the rest—an unlabeled residual category—constitute the majority of working theorists and work at an intermediate level of abstraction. Mathematical physicists are more distinct; they have specialized journals and may receive their training in departments of mathematics.[15] Most theorists dislike working directly with data from experiments, even if they may be dubious of the abilities of experimental physicists to do this. The education of experimental physicists has recently been catching up with the theoretical revolutions of the past few decades, and more experimenters are competent to begin the theoretical analysis of their results. One eminent theorist said: "I can do phenomenology; I am able to make the connection, but don't choose to. I would like to force experimenters to do their own phenomenology, and they can." Despite this, communication between theorists and experimenters is often difficult.

"There is a great language barrier" between theorists and experimenters, as one informant said. Although theorists may not be competent to understand details of the methods by which experimental results are obtained, they can usually appreciate the results themselves. Communication in the other direction is more difficult. Experimenters may attribute this difficulty to the values of theorists or to the differentiated role system. One, participating in a radio panel discussion,[16] argued that the theorist and experimenter were too widely separated and that many theorists were not really interested in the empirical world, but only in mathematics. (The solution he proposed for the problem was to create a third role between those of the theorist and the experimenter.)

Because of these barriers, leading experimenters and theorists seldom rely solely on written communication; face-to-face contact is necessary for effective communication. A Nobel Prize-winning experimenter said:

> When I was working on strange particles, I [had many contacts with theorists]. X was extremely interested in them and made some suggestions about experiments. Y, when he first came here, hung around all the time, and he eventually cooked up the best current theory on strange particles.

"X" in this quotation, when asked whether he could work without contacting experimenters, said: "It would be very hard. And no fun. It would be possible, but hard. For example, the Japanese theorists have a hard time." "Y" said that such contacts are:

> Absolutely necessary. . . . For one thing, I like to influence the experiments. I can do it one way by writing articles suggesting experiments, but it's simpler and quicker and more convincing to go and argue with someone and persuade him to do an experiment. Also, by talking with these people, one can get an idea of possible experiments and one can get an idea of new things one ought to calculate. . . . I find it easy to talk to [experimenters] about physical experiments. I have to keep asking them elementary questions about experimental possibilities, but they answer them readily, and we go on.

One theorist agreed that there were "communication barriers between experimenters and theorists," but suggested that this is "more often personality than physics." This should be interpreted as meaning that considerations of prestige and power are more important in causing strains between theorists and experimenters than are language barriers.

Role differentiation in physics has created (1) communication barriers, (2) resentment among those deprived of status, and (3) dissatisfactions in experimental research. Despite this,[17] the organization of physics in the past fifty years has been highly effective. The very presence of the strains indicates the integration of theorists and experimenters in the single discipline.[18] Experimenters can feel their status deprivation keenly because they are part of the same status system as theorists; i.e., *experimenters and theorists award recognition to one another* and thereby influence one another's behavior. They share the goals and values that make this kind of information-recognition exchange possible. Social integration has technical consequences. In recent history, theory and experiment have continually fed into each other; first theorists, then experimenters, have gone ahead. Theorists have made predictions—such as the existence of neutrinos and minus omega particles—later confirmed by experiment, and experimenters have made discoveries—such as pions and strange particles—which posed repeated challenges for theory. The lags between the two have been of short duration. One lag sometimes mentioned concerns the overthrow of parity, perhaps the most important "revolution" in physics in the last decade; but in this case less than six months elapsed between the suggestion by Lee and Yang that parity should not hold and the experimental verification of the prediction by Wu and her colleagues.[19]

As theories and data-gathering procedures become more complex in other disciplines, they, too, will become functionally differentiated. Theoretical roles are already coming to be differentiated in such fields bordering on physics as astronomy, physical chemistry, and biophysics. Such differentiation may have profound consequences for the ethos of science, consequences on which we can now only speculate. The establishment of theoretical elites especially responsible for fundamental and rigorous scientific theories may weaken the democratic spirit in science. If only a small core of specially trained scientists can understand and manipulate theory, the idea that the validity of theories rests partly on their intersubjective verifiability, their susceptibility to test by all qualified observers, will be even more limited. The idea of the "qualified observer" has always been something of a myth, since qualified observers were only to be found in the community of scientists. But the establishment of theoretical roles might mean that the scientific community itself would be divided between a clearly identifiable elite and a larger group which necessarily follows its lead. Such speculations as these are not supported by observation of the division of labor in physics. Despite changes in its social organization, physics has retained, not only its effectiveness, but the essentially democratic character of the scientific community.

NOTES

1. *The Positive Philosophy,* freely trans. and condensed by Harriet Martineau (London: John Chapman, 1853), II, 459 f. *et passim.* See also J. D. Bernal, *The Social Function of Science* (London: Routledge, 1939), pp. 113 f.
2. The usage of "segmentation" and "functional differentiation" here differs somewhat from that of Émile Durkheim, Talcott Parsons, and other sociologists. For a defense of this usage, see my doctoral dissertation, "Social Control in Modern Science" (University of California at Berkeley, 1963), pp. 426–428.
3. The differentiation of research groups into such roles as professional researcher, student, and technician was described in Chapter III. The point of the discussion was that, although the researcher was dependent on his assistants, he continued to make key decisions himself. That is, the role of the scientist was not differentiated into interdependent *professional* roles.
4. *Archives of Biochemistry and Biophysics,* 93 (June 1961).
5. "Aufnahmebedingungen," *Biochemische Zeitschrift,* 333 (1961).
6. *Biochemical and Biophysical Research Communications.*
7. See his popular summary of some of his work, "The Structure of the Hereditary Material," *Scientific American,* 191 (October 1954), 54–61.
8. In such relatively undeveloped fields as sociology, theoretical and field-research roles may appear to be differentiated. However, the interdependence of "theor-

ists" and field researchers is relatively low, and it is easy to move from one role to the other. Calling this "functional differentiation" is not justified.

9. Max Planck, *Scientific Autobiography and Other Papers* (New York: Philosophical Library, 1949), pp. 33 f.
10. *Quest: The Evolution of a Scientist* (New York: Doubleday Doran, 1941), p. 146.
11. Boston: Little, Brown, 1949. Wilson was himself trained as a physicist and probably wrote from experience. The novel is set in American universities in the period 1932–1946.
12. The hostility in the novel was linked with anti-Semitism. The same thing actually occurred in Nazi Germany. Some German experimenters—led by Lenard, an experimenter who believed that he had not received enough recognition for his work on X rays, although he did receive the Nobel Prize in 1905—seem to have interpreted "Jewish physics" to mean, not only Einstein's theories, but abstract theories in general. This approach was defeated, in part because of the enormous prestige of Werner Heisenberg, but not without a debilitating fight. See Samuel A. Goudsmit, *Alsos* (New York: Henry Schuman, 1947). Actually, the Nazis may have been correct in perceiving a correlation between "Jewishness" and theoretical physics, for Jews are more disproportionately represented among theorists than they are among experimenters. Thinking only of Americans, Nobel Prize-winners in experimental physics include R. A. Millikan, A. H. Compton, Carl Anderson, Owen Chamberlain, Edwin McMillan, and Donald Glaser, only the last of whom is Jewish, whereas leading American theorists include such men as Robert Oppenheimer, I. I. Rabi, Julian Schwinger, Richard Feynman, Murray Gell-Mann, John Bardeen, T. D. Lee, and C. N. Yang, most of whom are of Jewish descent.
13. Cf. Bernice T. Eiduson, *Scientists: Their Psychological World* (New York: Basic Books, 1962), pp. 135–139.
14. See Laura Fermi, *Atoms in the Family* (Chicago: University of Chicago Press, 1954); Emilio Segrè, "Biographical Introduction," *Enrico Fermi: Collected Papers* (Chicago: University of Chicago Press, 1962); and Jeremy Bernstein, "A Question of Parity," *New Yorker,* 38 (May 12, 1962), 49–104.
15. R. B. Lindsay, in William L. Schaaf, ed., *Mathematics: Our Great Heritage* (New York: Harper, 1948), p. 203, suggests that mathematical physics is a "genuine branch of mathematics, too often confused in popular parlance with theoretical physics." The theoretical physicists interviewed for this study did not agree.
16. On station KPFA, Berkeley, California, March 26, 1958.
17. And despite the disadvantages of working in formally organized groups; see Chapter III.
18. Parallels to this have been noted in other contexts; the existence of competitive strains is evidence of the importance of recognition in the behavior of scientists, and the existence of conflicts over right of access to laboratory facilities is evidence of the conformity of scientists to the norms of a larger scientific community.
19. J. Bernstein, *op. cit.* In the same article, Bernstein points out that experiments demonstrating the nonconservation of parity were performed as early as 1928—before the data could be made theoretically meaningful.

VI

THE CONDUCT OF DISPUTES

In the course of disciplinary differentiation, ideological differences sometimes lead to disputes between scientists about the merits of different goals. These disputes are symptomatic of social strains and are usually resolved when differentiation is completed. The claims of each of the opposing parties may eventually become valid and acceptable to the other. When the groups have become differentiated, both parties may adhere to their original goals and standards, but it will no longer be felt that they contradict each other. Consequently, segmentation and functional differentiation are examples of *logical evolution;* higher order social norms (the norms of science) are specified and differentiated for lower order structures.

Consider, now, cases where competing approaches to the same material are presented, such as contradictory theories purporting to explain the same data. Such contradictions cannot be resolved by social differentiation. One or more of the competing approaches must be mistaken, and the outcome of the dispute will be the transformation of the outlook of the discipline. This intellectual process may be associated with the process of *dialectical evolution,* a process that may involve pathological consequences for the organization of science. Opposing schools of thought may become so alienated from each other that information and recognition are exchanged only within a school.

From a slightly different point of view, the existence of opposing schools of thought reveals a possible strain between two central sets of norms in science. There are, first, norms giving the individual scientist liberty to accept or reject alternative approaches. Other norms provide that recognition and evaluation of scientists, even their right to be considered as scientists, should depend on competence and excellence. The problem is to apply this second set of norms without compromising the first.[1]

The Positivist Ideal

After interviewing many American university professors, Caplow and McGee reported:

> In the natural sciences, the question "Is there any particular professional viewpoint with which he is identified?" tended to be resented by the respondents, apparently because the physical scientist thinks of "viewpoints" as being polemical and believes that science has outgrown them. Note the unanimity in the following set of responses from one university's physics department.
>
> "I'm not aware of different viewpoints or schools of physics."
> "There are no schools in physics."
> "No, I don't. I never thought of physicists as belonging to schools."
> "That's not a meaningful question in science."[2]

This opinion is very common in the physical sciences, and it is often defended with arguments like those made by some philosophers. For example, an experimental physicist said:

> To me it seems that the psychological trends in physics are towards useful models. You notice the word 'model' is used quite a bit in the literature. . . . This reflects an emotional frame on the part of the modern physicist, that the truth is so difficult to obtain that the thing to do is to quit being upset by it and to try and obtain useful models to describe phenomena. For example, you will create a nuclear model which is very useful for certain types of data available but which may not be any good for other types of data, so you get another model for these other types of data. So you learn not to find this distasteful. Outsiders tend to feel that the world is understandable in some absolute sense, and I think we learn after awhile that it is not. . . . I would rather put it like this, that a theory in physics is not a religion, it's a policy. You don't want to be emotionally committed to a theory, it's just useful to indicate what kinds of experiments would be worth-while embarking on. If the experiments turn out to be completely different, this is nothing to feel outraged about, you simply proceed to change the theory.

A theoretical physicist described a similar point of view, one he personally did not share:

> Some people argue that the theoretician's role should be to present essentially mathematical conclusions. It doesn't affect his creativeness, it's just a question of style. That is, he should say, "If the following is true, then we

> can conclude so-and-so and this can be checked by experiment." They argue, then, that such an article is mistaken only when there is a mathematical error between the premises and conclusion. That is, if the whole thing turns out to be irrelevant, the premises wrong and the conclusions likewise, that's not a mistake, nothing to be ashamed of, nothing whatever. . . . Such people have no hesitation, when they are sure the mathematical work is sound, to publish. The premises say this, the conclusions say that, the mathematics is correct, there is nothing wrong with the article. Publish it at once. . . . But I cannot persuade myself to write in that style.

A physical chemist, who refused to commit himself to alternative theories regarding the chemical bond, was charged with accepting the idea that theories were merely convenient fictions, an idea much like that of the defenders of the Ptolemaic system in the sixteenth century. He replied:

> But there was really nothing wrong with the Ptolemaic theory. *That's right, but Copernicus was right, after all.* He was right in that he had a simpler explanation. You can explain the universe all right with the Ptolemaic theory, but things get very complicated. Things don't fall into a nice pattern.

In these cases, what might be called the "positivist ideal"[3] is more or less accepted: scientists feel there is *no reason* to become intensely involved in disputes. Hypotheses and theories are either meaningful or meaningless, either susceptible to empirical verification or not. There is therefore no justification for individual commitment to an unverified theory, or for the formation of schools of thought, and in the same way there is no justification for refusal to accept theories that meet accepted standards of validity.

Whatever the validity of the positivist ideal as a norm—and it is questionable—it is certainly false as a description of the behavior of scientists. Scientists do become committed to theories for which there is insufficient evidence to convince their colleagues, they do argue strenuously among themselves about the validity of opposed theories, and these arguments do become socially structured. There are "schools of thought" in physics as well as the other sciences. Scientists often act as if they did not accept the positivist ideal or as if its scope were limited.

LIMITATIONS OF THE POSITIVIST IDEAL

The positivist ideal is most applicable when conditions are right for a well-established consensus on methods and theories. Four such conditions can be specified.

(1) Intellectual scope of the disagreement. The positivist ideal will be most applicable when disagreements have a sharply restricted scope. An example would be contrary theories of the mechanism of particular organic-chemical reactions. Chemists can disagree about these while re-

maining in agreement about the general physical chemistry of reactions, the structure of organic compounds, and so forth. As the scope of disagreement widens, the positivist ideal becomes less applicable. For example, whether radiation was interpreted as waves or as particles once conditioned one's interpretation of a wide range of physical theory and experiment.

(2) Scope of the programmatic implications of the disagreement. Theoretical and methodological positions determine the direction of research and the design of particular experiments. When disagreements imply only a small difference or no difference at all in this respect, the positivist ideal will be applicable. For example, a particular disagreement might affect the design of only a single experiment or a small series of experiments; even if the experiments cannot be performed easily, disagreements about their design will not lead to serious disputes. On the other hand, some disagreements imply different programs, not simply for a few experiments, but for the research of a large portion of a discipline. The differences will concern the meaning of problems in general and the methods to be used in solving them. For example, whether or not a mathematician working in any of a large number of specialties accepts certain ideas relating to the concept of infinite sets may affect the design of his own work for years; it may also influence his choice of close colleagues and thereby lead to the formation of schools.

(3) Ease of making decisions. In most areas of research, scientists can be confident that a relatively few discoveries will lead to general agreement about the respective values of alternative approaches, and they can expect such discoveries in a relatively short period of time. In such cases scientists will concentrate on objective possibilities and avoid taking strong positions on what might be quickly shown to be invalid. Thomas S. Kuhn has suggested that the degree of quantification of a field is closely associated with the ease of making decisions and of achieving consensus:

> Probably for the same reasons that make them particularly effective in creating scientific crises, the comparison of numerical predictions, *where they have been available*, has proved particularly successful in bringing scientific controversies to a close. . . . Analytic, and in part statistical, research would show that physicists, as a group, have displayed since about 1840 a greater ability to concentrate their attention on a few key areas of research than have their colleagues in less completely quantified fields. In the same period, if I am right, physicists would prove to have been more successful than most other scientists in decreasing the length of controversies about scientific theories and in increasing the strength of the consensus that emerged from such controversies.[4]

When quantitative precision is not available, when disagreements are highly complex, or when disagreements involve standards of evaluating evidence, scientists will be uncertain about the possibility of convincing their opponents with any particular types of evidence. For example, a molecular biologist said:

> When you talk about molecular biology—or what I do, macro-molecular biology—this is an area which is rather hard to work in, and we can only collect a certain amount of data about [molecules], so there is an enormous discrepancy between the number of facts that you need to really specify the molecule and the number we have available. This is not true, for example, in carbohydrates and fields of biochemistry that deal with small molecules. So, since we have a small number of facts about complex problems, we tend to generalize, and this is the way we generate controversy. If one group has one set of facts, and another group has another, both of which are very restricted, they tend to draw up different pictures of the molecule.

In such instances scientists may rely more strongly on methods of persuasion that are not clearly "objective."

(4) Implications for textbook education. As long as disputes about subjects do not affect textbook education, many scientists, particularly those primarily oriented to teaching, will be able to stand aloof from them. When disputes come to have educational implications, either because of their revolutionary nature or their importance for applied fields, or because the disputes are essentially about teaching methods, this may no longer be possible. Most scientists must assume a position and will be brought into the conflict. For example, the subject of "interval estimation" or "fiducial intervals" is an essential part of the instruction of elementary statistics. The positions associated with Jerzy Neyman and R. A. Fisher are contradictory on this subject; there are texts offering each of these contradictory points of view; and most teachers of statistics must make some sort of decision.

(5) Degree of consensus. Even if none of the preceding conditions for the applicability of the positivist ideal holds, there may be a strong consensus in the discipline about the matters under dispute. One party to the dispute will be in a weak, minority position, for essentially nonsubstantive reasons. When it presents little threat to the consensus, it may be tolerated; open attacks on the position may be considered improper, and it will not be deemed important enough to require the familiarity of most scientists. (Those holding the minority position may form a school, but not one that is opposed by other schools.)

When the intellectual scope of the disagreement and its programmatic implications are restricted, when decisions can be made relatively easily and quickly, and when the dispute has no implications for textbook education, the positivist ideal will be most applicable. (Whether it is actually applied will depend on the institutionalization of the appropriate norms in the discipline.) On the other hand, when disputes have a wide intellectual and programmatic scope, when decisions are made with difficulty, and when the dispute affects textbook education, socially disruptive disputes will be most likely to occur.

TYPES OF DISAGREEMENT

Serious disputes may be initiated when a leading scientist perceives that particular discoveries or points of view imply extensive changes in the existing approaches of a discipline, and then goes on to attack either the innovation or the established procedures. Scientists may not perceive all the implications of innovations and may fail to recognize the possibilities of cleavage for long periods.[5] The eventual formation of schools of thought may be started either by exponents of the innovation or by its opponents. A wide range of innovations may initiate conflict between schools.

The most important, most conspicuous, and best known sources of substantive disputes are theoretical revolutions.[6] Revolutions are preceded by a state of crisis in which a generally accepted theory is shown to be inadequate for explaining certain critical data. Such crises are extraordinary events in developed sciences. "Normal science" is characterized by almost universal consensus on certain paradigms, "accepted examples of actual scientific practice—examples which include law, theory, application, and instrumentation together—[providing] models from which spring particular coherent traditions of scientific research."[7] No paradigm is perfectly consistent with the observations scientists are led to make; always some observations violate paradigm-induced expectations. In normal science, these may be given little attention or may be treated as "anomalies," and scientists will attempt to compel nature to conform to the paradigm by experimental or theoretical refinements consistent with it. The state of crisis arises when scientists persistently fail to account for anomalies with existing paradigms. Crises are resolved by theoretical revolutions involving more or less extensive destruction of previous paradigms and major shifts in the problems and techniques of normal science. Typically, the paradigms resulting from such revolutions successfully cope with previous anomalies, although they may abandon as scientific problems what previous paradigms considered problematical or explained. For example, after the revolution in chemistry associated with Lavoisier, chemists ceased to treat such qualities as color and taste as problems for their science, although previously chemists had considered these to be appropriate problems that had been partially solved.

Theoretical revolutions may have serious consequences for most of the research of an entire discipline, or of many disciplines. In restricted areas, revolutionary innovations may easily be recognized as superior to alternatives, but to determine the validity of the innovation for the entire range of subjects to which it applies will be more difficult. Many scientists will resist its application to areas in which it seems to lack empirical support. Other will attack it *in toto* because it seems to contradict well-established theories and methodological standards. Co-operative efforts may be made to disprove or prove the innovation, and for a time the discipline may include conflicting schools of thought. Examples of major revolutions in the sciences are familiar to everyone. The twentieth century has seen the revolutions introduced by the theories of relativity and of quantum mechanics in physics, and in the same period the life sciences have been revolutionized by the theories of evolution and Mendelian genetics.

Theoretical innovations need not be so spectacular as evolution, relativity, or quantum mechanics to deserve the label "revolutionary." In more restricted areas of science, scientists often make theoretical discoveries that imply extensive revision of accepted concepts and theories, as well as new and different directions for research. Because of the more restricted scope of such innovations, the involvement of entire disciplines is far less likely. For example, in the early nineteen fifties, the discovery of new particles led to a crisis in theoretical nuclear physics, one which is not yet completely resolved. As a solution to one aspect of the crisis, in the summer of 1956, Yang and Lee suggested that the law of the conservation of parity should be rejected for weak interactions.[8] Their suggestion was vigorously resisted by some physicists. Two informants mentioned this:

> [The law of conservation of parity] was apparently very well established. So much so that one of the experimenters, a man of great distinction, said, when a new experiment was proposed on the topic, that it was a complete waste of time, that these people shouldn't do such things, to check things which we know to be correct.

> One physicist had said previously, more or less publicly, that if Yang and Lee were correct he'd eat his hat.

Critical experiments were performed during the following winter, and overnight the physical community rejected the law. Consensus is not always achieved so easily. A biochemist, for example, might suggest that a particular complex set of reactions occurs in color vision, an hypothesis that differs from one generally accepted in his specialty, and adduce some evidence for his position; since testing hypotheses in this area is extremely difficult, it is possible for a dispute about the theory to be prolonged for years.[9]

Theoretical revolutions characterize only those sciences having well-developed empirical theories; there can be none without a general theory to be overthrown by the burden of accumulated empirical data. Corresponding to them in the formal sciences are disputes pertaining to the foundations of theories. Efforts to develop new foundations may be stimulated by the discovery of paradoxes in theories, a crisis similar to the crises that give rise to revolutionary theories in the empirical sciences. For example, the discovery of paradoxes in set theory, around the turn of the century, stimulated mathematical logicians to more strenuous efforts in the foundations of mathematics.[10] When opposing points of view develop with regard to foundations, they do not necessarily have much effect on existing theory. Opponents with respect to foundations might agree on the truth of most of the derived propositions held in the discipline, as well as on the general direction research should take. Formal scientists are seldom required to abandon large portions of their theories; instead they attempt to provide foundations to save as much as they can. It is true that a current school of mathematical logicians, the Intuitionists, argue that much of contemporary mathematics is based on unsound foundations and may be internally inconsistent, but this position has received little support. Similarly, some exponents of "subjective probability," a concept at the foundations of statistics, claim that it may have wide-ranging effects on the use of statistical methods.[11] The main reason such disputes lead to the formation of schools is the difficulty of coming to a decision, and this is directly related to the fact that the disputes have few consequences for most research scientists.

In the twentieth century, serious disputes about foundations have occurred in logic and statistics. With regard to the foundations of mathematics, three clearly opposed schools have developed: the logistic, represented by Russell and Whitehead; the formalist, represented by Hilbert; and the intuitionist, represented by Brouwer and Heyting.[12] In statistics, some current disputes seem to be of this type, including those between Fisher and Neyman,[13] along with the arguments about subjective probability.

In the same way that "minor revolutions" can be said to occur in empirical sciences, disputes occur with respect to the foundations of limited areas within the formal sciences. For example, there may be opposing schools with respect to the foundations of algebraic geometry or other mathematical specialties.

In both formal and empirical sciences, various techniques may be offered to aid in the solution of essentially the same kind of problem. Sometimes the differences imply different substantive positions, different views of reality. Sometimes scientists are unable to determine whether alternative techniques are consistent. When it is not clear which methods are superior, or what the criteria of superiority should be, the proponents of different methods may engage in prolonged disputes about the respective merits of each. Experimental physicists, for example, may argue about alternative methods for observing the effects of nuclear events. In mathematics, the followers of Sir William Rowan Hamilton sought to convince applied mathematicians of the superiority of his quaternions over various methods of vector analysis, which were presented as alternatives to it.[14] Molecular biologists may argue about the merits of different analytical techniques, and similar disputes occur in other sciences.

Disagreements about methods shade over into disagreements about the "style" of scientific work. Such disagreements are frequent, yet it is usually difficult to discover what the issues are, since they are inherently vague. An example of such a disagreement is that between "classical" and "modern" or "abstract" mathematicians. (Those who call themselves "modern" are usually called "abstract" by their classical opponents.) A graduate student described the disagreement this way:

> Mathematicians quarrel and hold intense views on their subject. For example, the modern approach, or axiomatism, which is associated with the name of Bourbaki,[15] is opposed by those who hold more classical views. . . . The dispute is partly a question of generations: the youngest people support axiomatism. Of course, now things have become established, and fights are not as necessary. The older generation might feel stronger about these things—the careers of many were hurt by the change, and this hurt may lead them to feel more strongly. The older mathematicians here are more likely to be classically oriented.

The difference between classical and modern mathematics concerns the desirable level of abstraction:

> Classical mathematicians tie things down to real lines, or to particular coordinate systems, and so forth. . . . Modern mathematicians talk about *general* properties. . . . These modern methods are quite powerful. . . . There is a tremendous difference here between undergraduate mathematics and graduate mathematics—the former is classical, the latter abstract.

The change in the level of abstraction has been characterized by some mathematicians as a "revolution" in their field; this assertion is, however, often denied by classical mathematicians.[16] Still other mathematicians are almost unaware that a disagreement exists; many were quite unclear about the meaning of "classical" and "modern." To some extent this lack of awareness may be the result of the dominance of the modern approach. In one university department:

> There is some little conflict between those with a modern background and those with a classical background. All signs point to a victory of those with a modern background. Classical mathematicians are held in great respect, however; no one would think of attacking them.

Many mathematicians assert that the difference is entirely pedagogical. Such an assertion takes some of the sting out of the conflict by reaffirming that mathematicians agree on what is "true," but bitter controversies can also extend to matters of instruction. In universities, these may be limited by norms of academic freedom, but there are no such limits within secondary instruction. The current revolution in the teaching of high school mathematics involves making it more "modern," and this has drawn sharp protests from classical mathematicians.[17]

One aspect of the dispute between the classical and modern style in mathematics concerns the general approach to the selection of problems, and this tends to overlap the goal conflict between pure and applied mathematicians. Mathematicians like Richard Courant and John von Neumann pointed out that an indefinitely large number of mathematical theories are possible, since mathematicians are free to propose new axioms and define new concepts. They contended that, because of this, it is difficult to find purely mathematical grounds for choosing to develop one mathematical system rather than another; mathematics alone cannot determine what results are "significant." Von Neumann warned:

> [A]t a great distance from its empirical source, or after much "abstract" inbreeding, a mathematical subject is in danger of degeneration. . . . Whenever this stage is reached, the only remedy seems to me to be the rejuvenating return to the source: the reinjection of more or less directly empirical ideas.[18]

Modern mathematicians respond to such arguments by saying that "mathematics is developed enough so that it can generate its own problems," and they are fond of pointing out examples of developments in pure mathematics, such as "imaginary numbers," which turn out much later to be of great practical importance.

William James accounted for the apparent lack of conflict over beliefs in science by saying that "options" in science were neither live, momentous, nor forced;[19] if scientists didn't agree, they could shrug their shoulders, go their own ways, and wait for another day. The positivist ideal is that this ought to be the case. But a wide range of situations has been described in which this ideal is inapplicable. Scientists are often faced with alternative hypotheses, theories, and techniques. The scientist must choose from among them in order to engage in a program of research; he cannot afford to wait for additional evidence before beginning his studies. And his choices are of great significance to him, not only because his reputation and career may depend on them, but because his own sense of identity is often tied up with the approaches he uses; his actions would be meaningless if the theories he accepted were false.

Scientists do tend to overemphasize the degree of consensus in science—the impossibility of serious disagreement; at the same time they usually find it possible to assert that the expression of disagreement is a necessary part of the advancement of science. Typical sentiments were expressed by Alexander von Humboldt in a public lecture in 1828:

> The discovery of the truth without difference of opinion is unattainable, because the truth, in its greatest extent, can never be recognized by all, and

> at the same time. . . . Whoever has called that a golden period, when differences of opinions, or, as some are accustomed to express it, the disputes of the learned, will be finished, has as imperfect a conception of the wants of science, and of its continued advancement, as a person who expects that the same opinions in geognosy, chemistry, or physiology, will be maintained for several centuries.[20]

Controversy stimulates discovery by drawing the attention of scientists to critical problems. A molecular biologist said of controversies in his field:

> [They] are often settled quickly. There are a great many people working in these areas. . . . The Biological Chemistry Society alone, which limits members to the associate professor level or higher, has 1,200 members. Work is done at a great rate, and many people are looking hard for something to do. If there are any unresolved problems they are quickly picked up.

Under some conditions, however, disagreements lead to alienation and the formation of schools, and disputes like this may have disorganizing consequences for science.

Disagreement and Alienation

Substantive disagreements do not usually lead to the alienation of scientists from one another;[21] when alienation does occur it is usually "personal," in the sense that other scientists do not become involved. Scientists often account for alienation in terms of "personalities." A theoretical physicist dismissed the topic in this way:

> There are certain pathological cases. . . . Most people don't take this too seriously. I know of only two cases. In the first, X versus Y, the men had different views and said hard words at meetings, but there was no malice involved. People were amused and the men were friends. It's interesting, but as far as I know the dispute was never settled; it was just sort of passed over. On the other hand, in one current case, a man is very strongly involved, and people fear for his mental balance.

Another theoretical physicist said: "Once in a while there are small men who feel their importance is bound up with a particular theory and refuse to face the facts against it. There is no point in trying to talk with them." To illustrate, he mentioned some arguments he had had with an eminent physicist, and concluded by saying that the other was "clearly irrationally attached" to a radical point of view. On the other hand, although Albert Einstein had strong and nonconformist opinions about matters of fundamental concern in physics:

> He remained a scientist. Once X and I developed a theory and took it over to Einstein. He looked it over and discussed its merits, and said in conclu-

> sion that, if the theory were correct, his theory of general relativity would be mistaken, and that this could happen any time; it would take a single experiment. We hadn't seen the incompatibility between our theory and general relativity until then.

Disagreements are likely to be exacerbated and lead to hostility when participants question each other's technical competence. To do so is perfectly natural: to a participant in a dispute, the easiest explanation of his inability to convince his opponents will seem to be deficiencies in their procedures. A molecular biologist, after pointing out that it was hard to reach agreement through published communications alone, said:

> It's very difficult to make up your mind unless you can go to a Gordon conference. Even then I was disappointed. I was chairman of a session and hoped to be able to resolve the conflict at that time. But people in both groups were so emotionally involved that they could not settle it. Under the circumstances, it was too difficult for an outsider to do so. *Do some of these emotional involvements occur because persons' levels of skills are challenged?* Right, definitely true. . . . Not really a matter of speculation but also of experimental skills. At one point in the disagreement a person reported on an experiment and another said [the inconsistency with other experiments] was "just a matter of technique."

Similarly, the well-known dispute in statistics between Karl Pearson and Sir Ronald Fisher with regard to the degrees of freedom for Pearson's chi-squared test arose because Pearson was wrong in his mathematics. As an informed statistician said, Fisher "was better equipped mathematically than Karl Pearson. . . . Pearson was wrong and lost."[22] Such occurrences poisoned the relations between Pearson and Fisher, and their antagonisms affected British and American statistics for many years.

Substantive disagreements are also exacerbated when they become linked with priority disputes. Apparently much of the hostility directed against the Bourbaki group stems not so much from their "revolutionary" efforts as from the fact that "people think that Bourbaki don't give credit rightly."[23]

> Most antagonisms come not from [substantive disputes] but from slights. One mathematician might feel his work has not been given proper credit by another, and this might lead to personal antagonism.

Similar disputes about priority were linked with the substantive disputes between Karl Pearson and Sir Ronald Fisher, and between the latter and Jerzy Neyman.[24]

Substantive disagreements alone cannot be a legitimate basis for one scientist's condemnation of another. Scientists reject the maxim "Love me, love my dogma." The moral component must be added to the substantive disagreement; it can arise from "unfair" attacks on a scientist's technical competence, from the "theft" of his ideas, from his own abuse of power, or from a conflict of goals. The last of these is frequently accompanied by substantive disputes. Some examples have been cited earlier, such as the disputes between molecular biologists and biologists with a "natural history" orientation. Similarly, the birth of modern statistics in Britain was linked with a bitter dispute between the supporters of the Mendelian theory of inheritance, led by Bateson, and the supporters of Galton's theory, led by Karl Pearson and W. F. R. Weldon.[25]

No matter how alienation is first generated, it always tends to develop and spread in a kind of vicious circle.[26] A scientist who feels his competence has been unfairly attacked is likely to respond by attacking the competence of his critics; a typical example was provided by an eminent scientist:

> We asked [an agency] for a grant for computations and eventually got it. Later on, I heard that X had opposed the request when asked about it and was strongly critical of our work. Then we looked into his work on related matters, not exactly the same thing, and found that he also makes mistakes. We wondered what to do about it but did nothing. Then his work was emulated by three other authors, so Y and I wrote two articles criticizing his approach, one for the *Proceedings of the National Academy of Sciences*[27] and for [another] journal. . . . We wouldn't have written the articles without his previous action. We had no difficulties in publishing the articles, since we were quite polite.

Similarly, a scientist who feels his opponents have stolen his ideas is likely to point this out in public, and scientists who feel their opponents abuse their power are likely to deviate in the same way. The vicious circle can continue to develop, and it may involve more and more scientists, with the final effect of dividing the discipline into a number of schools. Such vicious circles are most likely to develop when participants do not come face to face. As a theoretical physicist said:

> *Between* universities bitter conflicts can exist. For example, if a person's ideas are publicly ridiculed, strong antagonisms may arise. This isn't as likely to happen within a department. People exchange opinions at an early stage in research and this usually leads to a modification of them.

When strong divisions of opinion have been formed, departmental colleagues will tend to share the same views. Strong disagreements within departments may be viewed as almost intolerable, for, among other things, they become linked with competition for position, students, and research facilities. Personal hostility is most likely to result from substantive disagreements when they are linked with these types of competition. This may be one of the reasons for the generally noted difference between the intensity of disputes in European universities, especially before World War I, and contemporary American universities; competition for position in the former was far more intense than in the latter. Furthermore, the power and authority of the European professor probably resulted in greater homogeneity of opinions among those subject to him.

Alienation and Deviation

Scientific contributions regarded by their recipients as invalid are rejected by being either ignored or referred to only as "bad examples." This possibility motivates scientists to conform to methodological standards. When sections of the scientific community are alienated from one another, however, members of schools are likely to feel that their opponents fail to recognize their contributions because of the conclusions advanced rather than because of the methods used. (For example, a statistician who said Fisher "was obviously wrong" in his dispute with Neyman also asserted that "one of Fisher's students gave a scathing review of Neyman's [elementary text-] book. He closed by saying there was something good about it at least, namely the physical appearance and the type-setting.") When this occurs, scientists may reject their opponents' judgments—deny the legitimacy of these judgments as sanctions—and seek confirmation of the validity of their work from those who share their opinions. Scientists come to feel that those from whom they are alienated deviate from norms either by failing to recognize contributions properly or by not using appropriate methods, or both. When the legitimacy of sanctions is questioned, other types of deviation become more likely.

A symptom of this kind of breakdown of informal control is the difficulty members of opposing schools have in communicating with one another. This might be expected in the theoretical revolutions of the empirical sciences, since these revolutions involve changes in standards and the meaning of common concepts. It is more surprising to find it in the formal sciences, since they are devoted to logical clarity. But opponents of Sir Ronald Fisher's position in statistics, for example, often make such statements as: "I can't understand fiducial estimation. It seems like some kind of mysticism to me." The prolonged dispute about the foundations of mathematics gave rise to similar misunderstandings. An eyewitness of the debates in Germany in the later nineteen twenties between representatives of the intuitionist, formalist, and logicist positions said:

> Misunderstanding is easy because each point of view has developed its own terminology. Sometimes the same terms are used, sometimes terms which those of different points of view are liable to misinterpret. Communication is sometimes quite difficult. . . . Brouwer [the Dutch exponent of intuitionism] was especially difficult in that respect. . . . He was rather firm on his views and firm on the use of his terms. When we pressed him to say clearly how he meant to use these concepts he became impatient. . . . Brouwer was very polemical against Hilbert. Hilbert had great prestige, was

> the leading mathematician in Germany, and was at Göttingen. Brouwer would say, "Our difference is this: I assert this and you assert that I am against this," and so on. Then Hilbert would say, "No, this is not at all the difference." So not only did they differ on mathematics, they differed on the historical situation, what Hilbert's view was, what was possible, what the difference between the two was. . . . Brouwer got very temperamental. . . . He became quite emotional and indignant.

Most of the logicians interviewed for this study felt that, although the schools still existed, they were no longer as alienated from one another.[28] One, however, felt that his colleagues were mistaken. He asserted that intuitionism and formalism (the positions of Brouwer and Hilbert, respectively) were essentially similar and opposed to the dominant position, logicism, although the label "logicism" has been or should be abandoned since "it has been associated with half-baked popular expositions"; he went on to say:

> It's possible that a person feels so antagonistic to formalism-intuitionism that he can't tell his students about it. . . . X [an eminent logician] doesn't know much about intuitionism either; he thinks it is a kind of psychopathology. . . . I'm not an intuitionist and I can't understand what they mean. And no wonder, they don't seem to understand one another. You know Brouwer is the God of intuitionism and Heyting is his prophet, but, for example, Brouwer more or less repudiated Heyting's book, *Intuitionism*, the only lengthy exposition of the school's ideas.

Disagreement among logicians about the extent and intensity of the dispute reflects the segregation of the participants and their lack of communication.

When the system of informal control is weakened, the decisions of formal authorities necessarily seem arbitrary and capricious. Formal authorities must make decisions about appointments, promotions, and access to channels of communication on the basis of criteria some segments of the scientific community find illegitimate. For example, a logician was under pressure to accept the appointment of another logician from a different school in his department:

> I find X close to being incomprehensible. But I must make compromises for the sake of the program. Y is also an intuitionist and is also incomprehensible. . . . His retention may be a condition of X's coming. . . . I cannot judge Y's ability. . . . If I make one decision the program will fail, while if I make another I will be in with people I cannot understand.

Usually, however, authorities support those whom they feel are correct on disputed matters; a man's position on such matters may influence chances of receiving appointments. For example, with regard to the dispute in statistics between supporters of Fisher and Neyman, an informant said, "Maybe there is a restriction of mobility. For example, there is no mobility between X and Y universities. . . ." Another informant noted this case:

> [A man] wrote a number of articles on an issue, and the statisticians [at X university] didn't like them, and he probably couldn't get a job there because of them. But then, what better way is there of judging a man than by his published work?

One informant, tracing the dispute back to Fisher's arguments with Karl Pearson over different issues, asserted that power could also be used in other ways against opponents:

> Sir Ronald Fisher is a querulous individual who was also a revolutionary. [His revolution] started in 1915. . . . At that time Pearson was authoritarian; he was the editor of the only journal and the director of a unique laboratory. He published Fisher's first paper but thereafter tried to stop him at every turn. As a result, Fisher developed an inferiority complex. He divided the world into pro- and anti-Fisher segments. He resisted any thinking against his points of view. . . . [At first] *Biometrika* was the only journal and it was controlled by one [Karl Pearson] who was not a nice man. And Fisher had a spectacular struggle with Pearson. . . . He acquired power after Karl Pearson's death. . . . Fisher became a Fellow of the Royal Society, did a great deal of applied work, and helped others. As a result, up to now only members of Fisher's monastery can be elected to the Royal Society. This is one facet of his power; membership in the Royal Society is very important to English scientists.[29]

While this dispute in statistics is unusually spectacular and long-lived, the possibility of the apparent abuse of power always exists when segments of the scientific community become alienated from one another.[30]

If appointments and the opportunity to publish are based on scientists' opinions about disputed matters, and if the dispute lasts long enough, the contending parties tend to become isolated from one another. University departments and journals become associated with one or the other of the contending positions. Leaders of schools, feeling unable to persuade contending parties, tend to think in terms of organizational power. For example, a theoretical physicist who was identified by others as the leader of a school, when asked what the most important determinant of the beliefs of physicists is with regard to the theory responded: "Training is more important than literature and meetings. It is very difficult to get into this subject by reading. Partly this is because of its current lack of mathematical foundations." Earlier in the interview he had said, regarding supporters of his point of view:

> They are not now widely dispersed, but the number of workers is constantly growing. Two or three years ago this work was done in only two of three places. . . . Then these research associates, who really do the bulk of the work, began to accept permanent faculty appointments else-

where. . . . You cannot work effectively by yourself but must have others to talk to. A new professor usually insists on having like-minded colleagues at the institution, or the opportunity to bring in [postdoctoral] fellows. . . . *What about X University?* No. I don't know if that is accidental. Theoretical physicists there are mainly interested in other directions. The same is true of Y University. . . . X and Y universities do not go into it because of the tastes of their senior people, but they have no strong aversions. . . . We have one very conservative young theorist from X University—maybe he is our adversary because he comes from X. Z says that X University trains art critics instead of artists. This man is very critical and instinctively tends to distrust new ideas. He gives us controversy right here.

This scientist and his colleagues would certainly resent any allegations that he was involved in a power conflict or that he abused power; while rather candid about his views, he showed little hostility. Maybe he is less restrained among those of his colleagues who are sympathetic toward his views; a physicist in the same department said:

[views about the disputed issue were expressed] bluntly, in private. People will say "So and so is an idiot," "That thinking is ten years old," and so forth. . . . *Do you get into arguments about it?* Yes, friendly arguments. I get more out of them than those I speak to. In many ways they are almost like religious arguments. *In some fields one's position on a disputed issue is used to evaluate one's competence, so persons with some opinions can't get jobs. Is this the case here?* I don't know. It is in the general field of the interaction of personal and professional opinion. People feel quite strongly about these things.

Whether or not scientists involved in disputed issues are cynical or tend to abuse power, the situation tends to lead them to think in terms of power, for example, of placing those who support their views in the right positions.

This may take the form of the "secret society." Many such groups have been influential in the history of science, such as the "invisible college" of seventeenth-century England and the "scientific Lazzaroni"[31] of mid-nineteenth century America. While these groups were not associated with substantive scientific disputes, the Bourbaki group has been. Most mathematicians, including members of the group, would claim that its "secrecy" was adopted in jest or "to avoid a boringly long list of authors on the title page,"[32] but the group has been effective in many ways as a revolutionary conspiracy. One mathematician, in describing the change toward abstraction in mathematics, said:

Up to about 1920 there was classical mathematics plus a certain amount of abstract work. In the nineteen thirties and forties there was a strong reaction to this. Bourbaki had a strong influence on this. They were a group of strong personalities, strong mathematicians. Many of them taught at the

Institute of Advanced Studies and there influenced most leading U. S. mathematicians.

Many of the best U. S. mathematicians have taken postdoctoral fellowships at the Institute and have been influenced by the permanent members; similarly, Princeton and Chicago have had leading departments of mathematics for many years, and their appointment of members of the school undoubtedly helped it influence other mathematicians.

Another response to the conflict situation is to appeal for assistance to groups outside the scientific discipline. Scientists who feel they have not received justice from their colleagues, who deny the legitimacy of their colleagues' standards, are sometimes tempted to appeal to scientists in other disciplines or to nonscientists. In the dispute about evolution in the nineteenth century, for example, many opponents of the idea appealed to the public. Among them was Louis Agassiz:

> Beginning in December of 1861, Agassiz subordinated all his other activities to the cause of bringing about an informed public opinion through the popularization of knowledge. Once more he took up the role of lecturer and teacher of science. The fact that the American public was at this time exposed to the most dangerous of scientific and philosophical ideas—the concept of evolution—made it all the more imperative for him to serve the interests of education by teaching his version of the truth about nature. The decision to combat the idea of evolution through popular lecturing and writing was of primary significance. It meant that Agassiz abandoned his professional defense of classical natural history, for he never again published an objection to evolution in a professional journal or scientific monograph.[33]

Asa Gray, in a letter, denounced this enterprise of Agassiz and called him a "prince of charlatans."[34] This kind of appeal may introduce alien standards into disputes, thereby compromising the autonomy of the discipline.

The process appears clearest in cases marginal between genuine scholarship and quackery. For example, for a long period most serious scientists repudiated Mesmer and those who used his techniques.

> They shut their eyes to an important scientific discovery because they could not stomach the conditions of its demonstration. Mesmer was a nuisance. He was a propagandist and a demagogue, and, behold, the whole world had gone after him. . . . It is thus no wonder that the scientists repudiated him, and it is also no wonder that the use of hypnosis passed from the hands of scientists to charlatans for nearly half a century. This is the scientific dilemma that I am discussing: does science preserve its purity and thus retard its progress by shutting its eyes to partial truths, and does it thus sometimes cut off its nose to spite its face?[35]

This reaction might seem to be most common in marginal sciences like nineteenth-century psychology, but it is not unknown in the contemporary exact sciences. It seems likely that appeals to unqualified audiences induce anxiety among scientists and lead them to take retaliatory action. Thus, in the case of Immanuel Velikovsky, author of *Worlds in Collision*, it has been alleged that leading scientists abused their positions of influence and power to deny radical views a fair hearing—even going so far as pressuring publishers to cease publishing Velikovsky's books and having defenders of his positions dismissed from their jobs.[36]

Nonscientists sometimes become involved in such controversies even when contending parties do not appeal to them, because they feel some of their values are threatened; it is not always possible to isolate the independent effects of appeals by scientists to them.[37] This was true of the evolution controversy and the controversy about the Copernican theory.[38] Clearly, the involvement of nonspecialists exacerbates scientific controversies.[39] This may be true of the dispute in statistics between Fisher and Neyman. Both parties have addressed nonspecialists; the strong interest of nonspecialists in the matter, and their lack of technical skills with which to judge it, have probably prolonged the dispute. Most of the mathematical statisticians interviewed for this study were uninterested in it, and some of them asserted that most of those involved were applied statisticians.

The preceding discussion has indicated how scientific controversies lead to deviation from scientific norms: failure to recognize work properly weakens the informal organization of science; norms of independence are abridged when scientists abuse power to further their own ends; and the appeal to nonscientists threatens the autonomy of science. Another consequence of conflict may be to make the beliefs of individuals more rigid; beliefs usually become more rigid when an individual's informal status is felt to depend on them in part. In some respects these kinds of deviation are similar to those which accompany goal conflict in science.

GOAL CONFLICT AND SUBSTANTIVE DISPUTES

Both goal conflict and substantive disputes are characterized by the withholding of recognition for research contributions, in the former, because the contributions do not conform to the central goals of the discipline and, in the latter, because they do not conform to accepted versions of scientific truth. As a result, the decisions of formal authorities, which usually follow the informal award of recognition, must appear arbitrary. From the point of view of some scientists, the power of authorities over appointments and the selection of manuscripts for journals will appear to be abused. One possible reaction to such abuse of power is the appeal to higher authorities—the university officials or the larger public.

The strains resulting from goal conflict can be resolved by the formal differentiation of disciplines; this is impossible with regard to substantive disputes. Because the schools contend for the same goals—they claim to be working on much the same set of problems and to have identical jurisdictions—the problem area cannot be divided between them. When differentiation is not possible, other mechanisms of control must be developed. These involve the segregation of disputing parties and a permissiveness toward apparent deviants in the short run; these mechanisms are analogous to the primary adjustments noted in the case of goal conflict. The social control of scientific controversy centers on the suppression of alienation in its early stages, before vicious circles can develop.

The Control of Controversy

Scientists who hold opposing views are unlikely to become alienated from one another if their disagreement is concealed or belittled. The ideology of science stresses its unity; scientific truths are those upon which all informed observers can agree. Both those who resist apparently successful innovations and those who propose them are interested in finding common ground and will often claim that the disagreement is superficial. Many scientists find such disputes intolerable, and this is possibly part of the scientific personality.[40]

The positivist ideal can be considered one expression of the interest scientists and others have in denying the reality of fundamental disagreements. Some of the early expressions of this ideal by nonscientists were made in connection with their desires to accept both scientific findings and apparently contradictory nonscientific beliefs. For example, Osiander, a Lutheran bishop, wrote an anonymous preface[41] to Copernicus' *On the Revolutions of the Heavenly Orbs*, in which he stated: "These hypotheses need not be true or even probable; if they provide a calculus consistent with the observations, that alone is sufficient."[42] And he went on to assert that Copernicus' work contained many "absurdities." Thus, any disagreements about the heliocentric theory would merely concern the "methods" used by astronomers in making predictions, not fundamental truths, and there was no cause for intense commitment to it or to other theories.[43] The idea that scientific theories are merely convenient fictions to be used in making predictions or summarizing empirical observations is common in such positivist approaches; it makes it possible to disagree on particular theories while agreeing on more fundamental matters.[44]

Many contemporary scientists share the opinions of Bishop Osiander that theories are simply convenient fictions. This often takes the form of saying that apparent disagreements are not really scientific, but in some cases scientists will simply deny that a dispute exists.[45] For example, a theoretical physicist who was interviewed about a disagreement regarding alternative theories said:

> I don't know if you would call that a dispute. Theory A is newer and it gives hope of solving some problems which cannot be solved by theory B. There is a lot of dead wood in papers done in the past using theory B. . . . Nobody upholds [theory B].

His colleague in the office next door answered a question about prolonged disputes in theoretical physics affirmatively and said, "There are schools—that's the word to use. The center of one of the schools is [here]." And a third theorist in the department said:

> It has never been demonstrated that [theories A and B] conflict, although many feel that theory A is much better. . . . I would suspend judgment but would use the simpler theory [probably theory B] for computations. . . . Most statements made about theory A are exaggerated. Some parts of it have been shown untrue. A complete theory A has never really been formulated. I've heard the most exaggerated claims—that it would solve all the problems in quantum field physics. I think most of these claims will prove to be groundless. . . . I actually think some of my colleagues should be more rigorous.

Finally, an eminent theorist asserted that the difference was one regarding tools, not physical reality:

> I know that the theorists [working with theory A] have not yet got hold of any really new physical principles. . . . I think there is no sense in sort of making a mystique of [the elements of theory A], as some people do.

Physicists not only disagree, but they also disagree about the nature of their disagreement.

Similar examples could be provided from other sciences. Some mathematicians, for example, assert that none of their colleagues are emotionally involved in the disagreement between classical and modern mathematics, while other mathematicians will cite cases of disagreements. In the field of molecular biology, one scientist answered a question about "competing theories" in his field by saying:

> There are no big competing theories, nothing like corpuscular *versus* wave theories of light. There might be smaller things, but I can't think of any right now. The general picture which is emerging people generally agree upon. Every once in a while conflicting observations may come up.

He was given, as an example, a dispute mentioned in the *Scientific American*, which involved the biochemistry of color vision.[46] He replied, "Maybe that was to make the popularization more interesting. I don't know of any. People agree on the facts." Other molecular biologists, including a departmental colleague of the person just quoted, cited examples of rather long-lived controversies:

> Of course, there are a lot of bitter arguments going on. One of the most prominent at present is the theory of antibody formation, where Burnet got the Nobel Prize for his ideas. Most of us think along entirely different lines, based on experimental evidence.

> A dispute has gone on for about five years regarding oxidative phosphorylation. . . . This could probably go on for a good deal longer. It is debated quite openly at meetings of societies.

> I'm more aware of rather small points that are raised and about which people disagree. . . . A particular reaction has been found, and questions about the intermediates, the sequence of steps, what the mechanism is, these raise quite a lot of furor. People get excited about it. Some of them get very emotionally involved.

The denial of the existence of disputes by many scientists indicates a wish that disputes were not present, a wish to dissociate one's self from them. On the other hand, the admission of the existence of disputes by scientists means that other mechanisms—rationalization and social control—probably exist to control them.

Scientists who recognize the presence of controversies often attempt to define them as nonsubstantive; they assert that the controversies are not about *scientific truths* but about scientific techniques, scientific tastes, or philosophical truths. The definition of controversy as limited to methods has already been described; contemporary theoretical physicists often define controversies in much the same way that Bishop Osiander defined the geocentric-heliocentric controversy. This definition of controversy implies permissiveness, for the superiority of one of two techniques can only be decided on the basis of experience. For example, a young theoretical physicist said, regarding the dispute already described between the supporters of theory A and theory B:

> [Many physicists] point out that theory A has not yet had any outstanding successes. [This is] fairly true, but this is a controversial point. It didn't have the immediate overwhelming success of other physical theories, such as Bohr's theory of the atom or quantum mechanics, which explained a great many things when they were first presented. But it has had some success, if not yet any overwhelming success. . . . I think theory A is the only thing of promise for working on nucleons and mesons. I don't think anyone can predict its ultimate fate.

If disputes are only about methods, scientists can continue to agree about "facts," about "scientific truths," and they should be permissive about those who prefer alternative methods. Nevertheless, disputes about methods are scientific, they may be difficult to resolve, and they may lead to hostility and alienation among scientists. Other definitions of disputes, by calling them "nonscientific," remove them from the area of legitimate argument.

À CHACUN SON GOÛT

Formal scientists, not bound to produce theories that conform to empirical observations, are free to construct many alternative theories. Nevertheless, controversies arise. Some, such as controversies about the foundations of mathematics, are judged to be philosophical, not mathematical. Other disputes, about the level of abstraction in mathematics or the sources of mathematical problems, are judged to be matters of taste. For example, an analyst referred to the Bourbaki group by saying, among other things:

> They have a rather dogmatic point of view about the theory of integration in its most modern form, that it should be the way they look at it, whereas there are some other people who say that the theory of integration shouldn't be mixed up with topology, but should be quite separate from topology, and then later the two could be brought together for certain results you want. And the French school says no, integration finds its natural beginning in this area, this is where it starts. . . . *Are things argued?* . . . It's just a matter of taste, not a matter of scientific right or wrong.

Another mathematician described the difference between classical and modern mathematics by saying, "It is not the truth but the value one places on mathematical work. . . . It is largely an aesthetic judgment." An algebraist elaborated this point in greater detail:

> Most analysis is concerned with the real or complex number system, and many analysts feel abstractions are no good if they don't lead to results in this system. It is a matter of taste. I have graduate students who come to like algebra because it's a clean subject, it doesn't have lots of epsilons and deltas to worry about. It's a matter of taste. *In such disagreements, how do people attempt to persuade one another?* I never do this personally. I think a person should pursue his tastes. Even in high school some people prefer geometry, with its concreteness, others prefer algebra, with its abstractness and generality. . . . In talking with non-mathematicians, what bothers me most is their feeling that mathematics is cut and dried, simply a matter of finding things true or false. This isn't true. There is much room for taste and aesthetic appreciation in mathematics. In many ways mathematics is more of an art than a science. You might call it an exact art.[47]

De gustibus non disputandum est: This is a norm, not a matter of fact. It is possible to argue about matters of taste, as artists and chefs know, but it is not possible to argue "scientifically" about them. When an issue is defined as a matter of taste, it is defined as an issue about which it is improper for scientists to attempt to persuade one another. Such a definition is not "obvious," and to the layman as well as to some mathematicians such definitions may seem mistaken. For example, some approaches may be more fruitful than others in leading to new results that a wide variety of mathematicians can use in their research, and although the relative fruitfulness of approaches may be difficult to ascertain, it appears to be a matter of fact, not a matter of taste.

PHILOSOPHICAL ISSUES

Many of the topics I had judged to be controversial issues in science, on the basis of reading simplified treatments, were defined as "philosophical issues" by my informants, not scientific issues. Many, perhaps most, mathematicians define the controversy about the foundations of mathematics this way. When asked about it, or about such special aspects of it as set theory, they often made statements similar to these: "That was a philosophical, not a mathematical, quarrel. Point set theory is more philosophical than mathematical. *Are you interested in the matter?* No. I take a naive point of view of the concept of set." "This is not really mathematics but philosophy."

Physicists sometimes react in a similar fashion to attempts to renew interest in deterministic interpretations of quantum theory. In recent years the orthodox and indeterministic "Copenhagen" interpretation has come under sharp criticism from Louis de Broglie, the Nobel-Prize-winner, and David Bohm.[48] Only two of the thirteen theoretical physicists who were interviewed expressed much interest in the issue; one of them was an unusual man in many respects and the other a young man who said, "It is of no consequence to my own work and probably not to physical research in general. I'll bet I'm about the only one you've talked to who has been interested in the subject."

Some physicists said that the issue was essentially philosophical, and none of them spontaneously mentioned it as an important unresolved issue when asked if they could think of any such issues.

Scientists quite properly do not pay professional attention to problems that in principle cannot be answered by making empirical observations or providing logical or mathematical proofs. "Metaphysical" problems are not in the domain of science. However, scientists apparently define some issues as "philosophical" rather than scientific in order to avoid them and the possibility of controversy over them, regardless of the underlying nature of the problem. After Newton's theory of gravitation was accepted, scientists began to define the problem of how a force could act at a distance as "metaphysical," although it again became scientific with Einstein's theory of general relativity; and chemists after Lavoisier came to define the problem of explaining the "qualities" of matter such as color and taste as metaphysical, although such problems were scientifically resolved (in principle) with the development of quantum mechanics in the twentieth century.[49] Although the exclusion of metaphysical problems from science may not be in question, whether or not a particular problem is metaphysical may very well be questionable. The definition of problems as metaphysical is a social definition and not solely a logical one.

Saying that a problem was philosophical had a negative connotation for most of the scientists interviewed. The negative attitude went beyond particular issues; most scientists seem to be hostile or contemptuous of philosophy, or at least indifferent to it. They perhaps dislike philosophy in both its prescriptive role and its critical role.[50] Yet, in some cases, the typical activity of philosophers can also play a role in restraining scientific controversy. Philosophical criticism often involves specifying the dispute, reducing it to a matter of two or a few alternative propositions. By preventing the intellectual generalization of the dispute, the procedure makes it more difficult to infuse intellectual positions with value, and it makes the formation of conflicting schools correspondingly more difficult. It cannot be demonstrated here that philosophers have actually played such a role in disputes, although this type of analysis may be one reason why the controversy about the foundations of mathematics is less intense today than it was a generation ago; one informant thought so.

The behavior relevant to the control of controversy which has been discussed so far has been primarily individual behavior. Individual scientists define or redefine substantive controversies as nonexistent, as being about methods, or as nonscientific. If such definitions are maintained, scientists may withdraw from communication about the disputed issue. They can maintain a fiction of consensus, and the vicious circles of alienation that sometimes start from such disputes will be prevented. While the *effects* of defining away controversies can be said to help control controversy, it is more difficult to determine the *causes* of such behavior. (No typical patterns of characteristics associated with various individual reactions to controversy were discernible. For example, the two physicists quoted above, who occupied adjacent offices but gave sharply different evaluations of the dispute between adherents of "theory A and theory B," had relatively equal formal and informal status.) Perhaps scientists are subject to conflicting norms—on the one hand, norms of independence, impelling them to respect the judgments of their colleagues, and on the other, methodological standards, impelling them to be always critical and skeptical. In substantive controversy these norms conflict. The scientist who feels he deviates in the eyes of others—for example, by accepting a novel theory—can avoid their moral criticism by defining the matter as nonscientific, in much the same way that homosexuals attempt to avoid the condemnation of others in their community by defining their approach to sex as an aesthetic preference. However, more is involved in the denial of the scientific reality of controversy than the desire to resolve a conflict of norms. Scientists appear to have strong personal needs for unforced agreement from others and may avoid any evidence that such agreement cannot be obtained.

Another manifestation of this type of behavior is given in the way scientists perceive the history of science. Thomas Kuhn has pointed out that scientists persistently make the history of science look linear or cumulative—they perceive the scientists of the past working toward the goals of the present or the goals already achieved.[51] Textbooks, for example, play down the role of revolution in science. Newtonian physics is presented as a special case of relativity, leading naturally up to it, instead of including basically different concepts of mass, distance, and time. The caloric theory of heat is, if not ignored, treated as true in part and needing only to be complemented by the kinetic theory.[52] Chemistry texts often quote Robert Boyle's definition of the chemical "element," which is in fact quite close to the version accepted today. However, Boyle himself, in the *Sceptical Chemyst*, was skeptical, and the definition he is credited with was offered by him as a paraphrase of a commonly accepted view that he rejected.[53] In general the presentation of scientific history by scientists dismisses controversy as the result of individual idiosyncrasy and error, of "personality." (Sometimes scientists may even extend the reinterpretation of history to their own work. The English chemist John Dalton, for example, wrote autobiographical accounts of his discovery of the principal of combining proportions in which it appears that he was interested from an early date in the problems he later solved. Actually those problems only occurred to him late in his investigations, at about the same time as their conclusions.[54]) Scientists not only have distorted perceptions about the degree of consensus in the contemporary scientific community, they have also distorted perceptions about the unity of present and past research. The goals and concepts of earlier scientists are made to seem consistent with present goals and concepts. This sense of working in an enduring tradition, firmly supported by a present consensus, makes the scientist's work seem meaningful and valuable.

The tendency of scientists to withdraw from controversial situations is the primary reason that such situations do not usually lead to personal alienation and the formation of schools. The nature of the withdrawal also implies that the scientist who does attack another's point of view deviates from norms; he acts unscientifically by attempting to introduce matters of taste or philosophy into scientific discussions. Such scientists will probably face the informal disapproval of their colleagues. There are also more formal sanctions against such behavior.

THE SOCIAL CONTROL OF CONTROVERSY

The authorities who control access to scientific channels of communication—principally editors of journals and their referees—can prevent polemics by refusing to publish polemical papers. Usually this rule is implicit, and those who submit papers may be mistaken in believing that editors adhere to it. (Authors will also hesitate to express themselves polemically for fear their papers will fall into the hands of a hostile referee.) Some editors may make their rules explicit. The editors of *Biochemische Zeitschrift,* in 1961, stated in their "Conditions for Acceptance" that, "Polemics are to be avoided, although short rectifications regarding the facts of the case are permissible."[55] Other editors make themselves clear by rejecting polemical work. One scientist said:

> I've been an editor of the biochemical journals, and such an article would never be accepted. They are accepted—at least as letters to the editor—in, say, *Science.* But you wouldn't find them in a journal such as the *Journal of Biological Chemistry. Why the difference?* Because the one is a research publication, and the other is general, more of a scientific newspaper or magazine.

In the meetings of scientific societies, most papers are also refereed, and in any case those who contribute papers seldom have time to present any extensive arguments against another point of view. Sometimes the chairman of a meeting can prevent the expression of controversy. A biochemist said:

> You can always stop it if you're chairman. . . . I stopped a pretty hot argument at a recent meeting of the National Academy of Allergy. . . . Two groups from the same university were at each other's throats. It got to be almost personal, so I had to stop it.

Biological and chemical periodicals also discourage the expression of controversy in the course of discouraging purely "theoretical" work. (See Chapter IV. Possibly the control of controversy is a major cause of the discouragement of purely theoretical work.) As long as arguments are confined to a narrow range of experimental data, few scientists can become involved. Such prohibitions do not completely exclude controversies from scientific communication channels, but they segregate them severely. Biological and chemical periodicals do publish "review articles" by eminent scientists, and their associated societies often sponsor symposia, which are later published and in which eminent scientists participate. In review articles and symposia, "speculation" and polemical expressions are usually permitted. In effect, then, these sciences segregate controversy. Outside of informal conversations, most scientists do not become involved actively in them, and the controversies are publicly expressed only by eminent men on special occasions.

Controversy is also segregated socially in other sciences. Thus, when disputes about the foundations of mathematics are defined as "philosophical," participation in them tends to be restricted to specialists of a certain sort and other mathematicians do not participate. In physics, theory is the province of a differentiated subgroup, and experimental physicists are unlikely to be intensely involved in theoretical controversies.[56] Whether the expression of controversy is restricted to a scientific elite or to a group with special training and skills, the segregation reduces the likelihood that vicious circles of alienation will occur or that schools will be formed. Most scientists will work on problems about which there is consensus.

DYSFUNCTIONS OF CONTROL

Behavior leading to the containment of controversy may have undesirable consequences. The social segregation of controversy may mean that some scientists remain unfamiliar with points of view that might benefit their own research. In a previous chapter it was noted that logicians often feel that mathematics would be improved if more mathematicians were familiar with the foundations of mathematics. A molecular biologist felt that the tendency of authors to ignore contrary points of view impedes research and actually makes it more difficult to come to a decision.

> In a particular issue, I'll be interested in seeing how it comes out. While I'm not on the editorial board for the *Biophysical Journal*, I was sent these papers for review, and one of my major criticisms was that they do not attempt to treat the data supporting the opposite position. I felt the authors should at least mention the papers supporting the opposite results and say, "We don't get them." I think this is what should be done, yet I have sympathy with the author of this one, because his opponent's papers are published in the *Proceedings of the National Academy of Sciences*, which are unrefereed, so he doesn't have anyone telling him to do this.

If many scientists are unaware of significant unsettled issues, or are not made aware of them, the result may be a poor allocation of effort among research problems. In effect, the social control of research is relaxed by withholding communication on some topics. (This relaxation of control is similar to the primary adjustments to goal conflict and has similar effects.) These are some of the costs involved in the prevention of serious disputes in science. To some extent such costs may be reduced. But some costs are the result of tensions inherent in the scientific enterprise. Scientists seek to formulate existential propositions in such a way that all those with access to the relevant data will agree about their validity. This enterprise cannot exist unless each scientist is free to accept or reject propositions according to his independent judgment of their correspondence with scientific standards. On the other hand, scientists are obliged to act to maintain these standards—to expose gaps in theories, criticize faulty methods, and prevent charlatans from abusing scientific communication channels. They must act to control the scientific activities of others. In science as elsewhere, the tensions between freedom and control are, in the final analysis, unresolvable in the short run.

The Resolution of Conflict

Scientific controversies are usually short and are resolved by new discoveries. The discoveries may convince exponents of one of the competing schools of thought, or they may be genuine syntheses of opposing viewpoints. For example, Niels Bohr's principle of complementarity settled the centuries-long opposition of wave and particular theories of light by incorporating both of them in a new theory.

Scientists who have been mistaken rarely recant in public.[57] One might expect such behavior to be more common, for some of the reasons impelling people in other areas of life to recant are also present in science: the scientist's sense of identity is often linked with his beliefs, so that fundamental changes in beliefs imply changes in self-conception. In such situations, public recantations, by inducing confirmation of the change from others, reaffirm the new sense of identity for the one who recants. But apparently this psychological aspect of recantation is less important than the sociological aspect, that recantation destroys old group loyalties and affirms new ones. The norms of independence and universalism in science make recantation unnecessary, for scientists are expected to be committed to scientific norms, not so much to the scientific community or segments of it, and expressions of loyalty to such communities are not required. Generally scientists who have publicly opposed an innovation keep quiet or attempt to show that they did not really oppose the truth of the innovation. (Few written confessions of their errors by men who had opposed innovations can be found, and the interviews conducted for this study didn't lead to any such; e.g., no physicists mentioned being skeptical about the nonconservation of parity before it was demonstrated, although some of them did mention *other* physicists who had expressed skepticism.) For example, in 1903 Pearson and Weldon published a paper arguing that biometrics and the Mendelian theory were not incompatible, although only shortly before this Pearson had argued that the nature of variation in the germ cells was identical to that of other bodily organs.[58] The defeated parties in disputes usually withdraw as gracefully as they can and learn to live with the dominant point of view. This is easily done, for most scientific assertions are eventually replaced.

Sometimes there are no discoveries that lead to a reconciliation of opponents; none of the evidence necessarily leads to conviction. But even if most *individuals* may remain steadfast in their views, the scientific *community* may develop a new consensus. As Auguste Comte wrote:

> There is no denying that our social progression rests upon death. . . . It is enough to imagine [the duration of human life] lengthened tenfold only, its respective periods preserving their present proportions. If the general

> constitution of the brain remained the same as now, there must be a retardation, though we know not how great, in our social development: for the perpetual conflict which goes on between the conservative instinct that belongs to age and the innovating instinct which distinguishes youth would be much more favourable than now to the former.[59]

These sentiments have been echoed by a number of scientific revolutionaries. Charles Darwin wrote, in the Conclusion to *The Origin of Species:*

> Although I am fully convinced of the truth of the views given in this volume under the form of an abstract, I by no means expect to convince experienced naturalists whose minds are stocked with a multitude of facts all viewed, during the long course of years, from a point of view directly opposite to mine. . . . A few naturalists, endowed with much flexibility of mind, and who have already begun to doubt the immutability of species, may be influenced by this volume; but I look with confidence to the future,—to young and rising naturalists, who will be able to view both sides of the question with impartiality.[60]

The same pessimism with regard to convincing his mature opponents by rational argument was expressed by Max Planck:

> Boltzmann eventually triumphed in the fight against Ostwald and the adherents of Energetics, as it had been self-evident to me that he would. . . . The basic difference between the conduction of heat and a purely mechanical process became universally recognized. This experience also gave me an opportunity to learn a fact—a remarkable one, in my opinion: A new scientific truth does not triumph by convincing its opponents and making them see the light, but rather because its opponents eventually die, and a new generation grows up that is familiar with it.[61]

It would be unwise to take these statements literally; after all, Darwin did convince such persons as Asa Gray, Charles Lyell, and J. D. Hooker, and Ostwald and Mach were eventually persuaded to accept the atomic theory. It would be just as unwise to accept the statements by Darwin and Planck without question as it would be to accept the statements of Robespierre and Lenin about their opponents in revolutions of another kind.

Nevertheless, there may be some truth in such assertions. Young scientists may find it easier to accept new views than old scientists, who may be more strongly committed to the earlier views. Although young scientists have been subjected to an education in which they are indoctrinated with accepted theories and seldom given any arguments against them, the commitments they make may be superficial. Firm commitments to a theory may be achieved only by those who have used it to account for things previously inexplicable, who have experienced the range of its power and the difficulties of subjecting it to test. Given a crisis or an innovation, the older generation may firmly believe in the possibility of reconciling it with the established theory. Younger scientists, on the other hand, will perceive most clearly the incompatibility of innovations and existing theory.[62]

Thus, the very wealth of knowledge and manipulatory skill of the experienced scientist makes it harder for him to accept innovations. Younger scientists do not have this "trained incapacity," at least not to the same extent. They will tend to take the "easy way" out of a problem, and this often means acceptance of the innovation. This is perhaps made clearest in disputes about methods. A person with much experience in the use of, say, quaternions, may be as adept at using them to solve physical problems as others are in using vector analysis. Differences in the methods are seen most readily by young men. (Taking a Machist approach, in which all science is considered in terms of relative "easiness" of method, this point can be immediately extended to foundations disputes and theoretical revolutions.)

The social positions of young scientists may also predispose them to accept innovations. They are not necessarily bound into a network of interpersonal relations that strengthens support of existing approaches. Since they are usually insecure in their positions, they feel more strongly the necessity of quickly producing important results. Furthermore, they are in a position that makes it convenient or necessary for them to select a new special field. All these things will tend to make them interested in innovations: innovations offer the promise of producing important results in areas in which there is less competition. In the same way, we might expect well-trained but marginal scientists, even if they are older, to be attracted to innovations—for example, Jews or orientals in places where discrimination exists.

There is therefore some plausibility in the hypothesis that many substantive disputes in science are also generational disputes—disputes whose outcome is as certain as death. It is difficult to substantiate this with much hard data. Hunting for isolated examples[63] without also hunting for contrary cases is of little value. While some mathematical informants felt that "classical" opponents of "modern" mathematics tended to be older men, some physicists asserted that the older men were more sympathetic to innovations. A theoretical physicist who is a leading exponent of an innovation asserted:

> Maybe there is even an inverse correlation with age. The only more radical group than our own is Heisenberg's. . . . They are on the lunatic fringe in most persons' eyes. But the younger Germans are very conservative. Pauli, who died a couple of years ago, was very interested in our work. Maybe older people, who lived through the invention of quantum mechanics, are more ready to accept new ideas, while younger people have lived through an age of relative stability in physics, from about 1928 to the present.

Thus, even if some disputes are generational, they need not be simply "innovative youth" versus "conservative age." Rather, the outlook of a generation is strongly influenced by events occurring when its members embark upon their careers.[64] Age may be more radical than youth.

There are undoubtedly characteristics besides age that induce scientists to resist or accept innovations. It would be interesting to discover the possible relations between resistance to innovation and social status in the larger society, although it would probably be more valuable to study the effects of status within the scientific community. But whatever the importance of other characteristics, age has the overriding importance of deciding the fate of the controversy, for the consensus established among succeeding generations is defined as scientific truth.

The Conditions of Tolerance

The manner in which scientists disagree can be held up as a model to the larger society. As Bronowski writes:

> Theirs is the power of virtue. They do not make wild claims, they do not cheat, they do not try to persuade at any cost, they appeal neither to prejudice nor to authority, they are often frank about their ignorance, their disputes are fairly decorous, they do not confuse what is being argued with race, politics, sex or age, they listen patiently to the young and to the old who both know everything. These are the general virtues of scholarship. . . . [The content of science has changed radically.] Yet the society of scientists has survived these changes without a revolution and honors the men whose beliefs it no longer shares. No one has been shot or exiled or convicted of perjury; no one has recanted abjectly at a trial before his colleagues. The whole structure of science has changed, and no one has been either disgraced or deposed.[65]

As he notes, this is necessary for science, for scientific truths can be achieved only by free actors.

Science also has other characteristics that make it possible. The scientific community is autonomous in the larger society. Scientific theories are only rarely, and then pathologically, ideologies to justify the power of groups. Within the scientific community, groups have ideologies, but, again, specific theories are only remotely linked to such ideologies. Thus changes in the content of science do not usually imply changes in social structures. Almost the only issue in controversies about scientific theories is the truth; this is a goal scientists find worth fighting for, yet it is a goal that can be achieved with only a restricted set of means.

The scientific community has a long temporal perspective. It seldom needs to make collective decisions on the basis of limited information, although its individual members must. The community can wait, it can afford to be tolerant.

The scientific community has usually succeeded by waiting. Eventually, discoveries are usually made that win the unforced assent of all its members. When such discoveries are not made, the scientific community may circumvent the problems giving rise to disagreement—scientists may be able to show that the problems dividing their community are not genuine, are not part of its concern.

In addition to these commonly known attributes of science, its autonomy, the necessary independence of its practitioners, and its long temporal perspective, the scientific community has other procedures for preventing the development of disorganizing disputes. When serious conflict threatens to develop, it is neutralized by segregation, by the temporary separation, voluntary or enforced, of disputing parties. These procedures have been described in the preceding pages. They are not always successful—warring schools have disorganized science—and they do have their own disadvantages. In most cases they have reinforced the other attributes of science to prevent the development of disorganizing conflict.

NOTES

1. Cf. J. D. Bernal, *The Social Function of Science* (London: Routledge, 1939), pp. 298 f.
2. Theodore Caplow and Reece McGee, *The Academic Marketplace* (New York: Basic Books, 1958), p. 88.
3. Although the term "positivism" as it is usually used aptly characterizes this point of view, it should be distinguished from "logical positivism," a more sophisticated philosophical position.
4. "The Function of Measurement in Modern Physical Science," *Isis*, 52 (1961), 185, 190.
5. Max Planck, for example, said he made attempts to fit his "elementary quantum of action into the classical theory . . . for a number of years, and they cost me a great deal of effort." *Scientific Autobiography and Other Papers* (New York: Philosophical Library, 1949), pp. 44 f.
6. The ensuing paragraphs draw heavily from Thomas S. Kuhn, *The Structure of Scientific Revolutions* (Chicago: University of Chicago Press, 1962).
7. *Ibid.*, p. 10.
8. For a simple treatment see Jeremy Bernstein, "A Question of Parity," *New Yorker*, 38 (May 12, 1962), 49–104.
9. An informant maintained a theory about viruses which he thought was fundamentally different from accepted theories and for which he was collecting evidence. When asked whether his point of view had given rise to any controversy, he replied, "I haven't presented enough of it as a theory, so there are no targets for them to shoot at." Such reluctance is typical.
10. Salomon Bochner has suggested: "In fact, the first foundation crisis was identified, in substance rather than in name, in twentieth century mathematics itself [set

theory], and past crises were then uncovered in the wake of this one." "Revolutions in Physics and Crises in Mathematics," *Science*, 141 (August 1963), 408–411. He suggests that the "crisis" with respect to the infinitesimal calculus, in the eighteenth century, was only recognized after a solution had been provided, and also that the ancient Greeks probably did not recognize that irrational roots provided a crisis for their mathematics.

11. See, e.g., L. J. Savage, *The Foundations of Statistics* (New York: Wiley, 1954).
12. See Max Black, *The Nature of Mathematics* (New York: Humanities Press, 1950).
13. See Lancelot Hogben, *Statistical Theory* (London: George Allen and Unwin, 1957); and G. A. Barnard, "Significance Tests for 2 × 2 Tables," *Biometrika*, 34 (1947), 168 f.
14. See Eric T. Bell, *The Development of Mathematics* (2nd ed.; New York: McGraw-Hill, 1945), pp. 206–211.
15. Nicolas Bourbaki was the *nom de plume* of a group of French mathematicians who set out to systematize and revolutionize mathematics over the last thirty or forty years. Their works run to at least twenty volumes. See Paul R. Halmos, "Nicolas Bourbaki," *Scientific American*, 196 (May 1957), pp. 88–99.
16. See Benjamin DeMott, "The Math Wars," *The American Scholar*, 31 (1962), 296–310, for a report on the controversy.
17. See *loc. cit.* and E. P. Rosenbaum, "The Teaching of Elementary Mathematics," *Scientific American*, 198 (May 1958), 64–73.
18. John von Neumann, in James R. Newman, ed., *The World of Mathematics* (New York: Simon and Schuster, 1956), Vol. 4, p. 2063.
19. "The Will to Believe," in *Selected Papers in Philosophy* (New York: Dutton, 1929), p. 115.
20. Quoted in Charles Babbage, *Reflections on the Decline of Science in England and Some of Its Causes* (London: B. Fellowes, 1830), p. 219. See also Babbage's own expression of this point of view, *ibid.*, p. 59.
21. The ensuing illustrative material therefore represents highly atypical behavior. The relative paucity of data on *social* conflict associated with substantive disputes in science means that much of the following discussion must be rather speculative. Bochner, *loc. cit.*, has noted the relative absence of social conflict in the major theoretical revolution of this century, that of quantum theory.
22. This is admitted by his son; E. S. Pearson, "Karl Pearson: Some Aspects of His Life and Work," *Biometrika*, 29 (1938), 212 f.
23. Halmos, *loc. cit.*, has noted that many exercises in the Bourbaki volumes are taken without citation from the work of others. This is likely to be especially aggravating because the "exercise" in the Bourbaki work may have been presented originally as a serious theorem, and, according to some mathematicians, may be so difficult and important that it should be treated as a theorem rather than an exercise.
24. See also Robert K. Merton on Freud's reaction to the suggestion that what was good in his work was taken from Janet: "Priorities in Scientific Discovery," *American Sociological Review*, 22 (1957), 653. The dispute about the relative merits of Newton's fluxions and Leibniz' calculus, with its disastrous consequences for British mathematics, was exacerbated by the priority dispute between the supporters of Newton and Leibniz over the discovery of the general method. See Eric T. Bell, *Men of Mathematics* (New York: Simon and Schuster, 1937), pp. 113 f.

25. See the biographical notes on Bateson in J. G. Crowther, *British Scientists of the Twentieth Century* (London: Routledge and Kegan Paul, 1952), ch. VI; E. S. Pearson, *op. cit.*, 28 (1936), 227–232; and Hogben, *op. cit.*, pp. 233 ff.; 250 f.; 287 f.
26. See the theoretical analysis in Talcott Parsons, *The Social System* (Glencoe, Ill.: Free Press, 1951), ch. 7.
27. This journal publishes papers submitted by members of the N.A.S. without passing them to referees, a procedure shared with the *Proceedings of the Royal Society* and some other journals. It is easier to publish polemical works in such journals.
28. Cf. A. Heyting, a leading intuitionist, "After Thirty Years," in Ernest Nagel, Patrick Suppes, and Alfred Tarski, eds., *Logic, Methodology and Philosophy of Science* (Stanford: Stanford University Press, 1962), 194–197. "The spirit of peaceful cooperation has gained the victory over that of ruthless contest." At the Congress at which this paper was delivered, however, there was some (good-natured?) heckling in response to it.
29. For further expressions of the Fisher-Neyman dispute see the following papers by Ronald A. Fisher: "Statistical Methods and Scientific Induction," *Journal of the Royal Statistical Society*, series B, 17 (1955), 69–78; "Scientific Thought and the Refinement of Human Reasoning," *Journal of the Operational Research Society of Japan*, 3 (1960), 1–10; and *Contributions to Mathematical Statistics* (New York: Wiley, 1950), paper 35. See also Jerzy Neyman's summary, "Silver Jubilee of My Dispute with Fisher," *Journal of the Operational Research Society of Japan*, 3 (1961), 145–154. Fisher's rather bitter articles against recently deceased opponents—against Karl Pearson (in *Contributions, op. cit.*, paper no. 29) and against Abraham Wald (the first paper cited above)—probably alienated his opponents still further.
30. The misfortunes of such revolutionaries as Semmelweiss are well known. Autobiographies of scientists sometimes mention the abuse of power; see, for example, Richard B. Goldschmidt, *In and Out of the Ivory Tower* (Seattle: University of Washington Press, 1960), pp. 60 f.; and Norbert Wiener, *Ex-Prodigy* (New York: Simon and Schuster, 1953), p. 208. Many scientists decry the resistance to innovation by scientific authorities; see, e.g., C. D. Darlington, "The Conflict of Science and Society," *Bulletin of the Atomic Scientists*, 7 (1951), 9–12; and, for the social sciences, C. Wright Mills, *The Sociological Imagination* (New York: Oxford University Press, 1959), pp. 108 ff. Bernard Barber has analyzed the resistance by scientists to innovations and provides some examples of appointments having been withheld from, and access to journals denied to, nonconformists; see "Resistance by Scientists to Scientific Discovery," *Science*, 134 (1961), 596–602. Logan Wilson also mentions the problem; *The Academic Man* (New York: Oxford University Press, 1942), pp. 208 ff. Some of the cases given by David L. Watson, *Scientists Are Human* (London: Watts and Co., 1938), pp. 56–67, and J. R. Kantor, *The Logic of Modern Science* (Bloomington, Ind.: Principia Press, 1953), pp. 52–58, seem to be of doubtful validity. Kantor, for example, refers to the Royal Society's refusal to print Benjamin Franklin's work on electricity, when in actual fact Franklin was feted by that group; see I. B. Cohen, *Franklin and Newton* (Philadelphia: American Philosophical Society, 1956).
31. See A. Hunter Dupree, *Science in the Federal Government* (Cambridge: Harvard University Press, 1957), pp. 135 f.; 138; 142.
32. Halmos, *loc. cit.*

33. Edward Lurie, *Louis Agassiz: A Life in Science* (Chicago: University of Chicago Press, 1960), p. 306.
34. *Ibid.*, p. 310.
35. Edwin G. Boring, "The Psychology of Controversy," *History, Psychology, and Science: Selected Papers* (New York: Wiley, 1963), p. 71.
36. See "The Politics of Science and Dr. Velikovsky," special issue of *The American Behavioral Scientist*, 7 (1963), 3–68. The papers in this issue are somewhat marred by the authors' strong beliefs in the validity of Velikovsky's theories and their consequent difficulty in viewing the behavior of his opponents objectively.
37. It is also difficult to define "scientist" or "specialist" in this context. For example, Hans Freudenthal reports "public" reactions to the discovery of non-Euclidean geometries (first published in the eighteen thirties but exciting more interest among mathematicians only in the eighteen sixties) by saying, "The Boeotians do not show up before the middle of the seventies—conceited people who claimed to prove that Gauss, Riemann, and Helmholtz were blockheads. If you witnessed the struggle against Einstein in the twenties, you may have some idea of this amusing kind of literature." ("The Main Trends in the Foundations of Geometry in the 19th Century," in Nagel, Suppes, and Tarski, *op. cit.*, p. 613.) The implication is that these "conceited people" do not count as "scientists." Yet Freudenthal includes the noted logician Frege among them (p. 618), and he points out that Carl Friedrich Gauss was concerned about their reactions—he never published a single word about his discovery of non-Euclidean geometry, "fearing, he said, the clamour of the Boeotians" (p. 613).
38. It is possible that Galileo's popular writings contributed to making the Copernican theory a religious issue. This is alleged by Arthur Koestler, *The Watershed* (Garden City, N.Y.: Doubleday Anchor, 1960), pp. 176 f.; 183 f. Religious and scientific issues were also linked in nineteenth- and twentieth-century France; efforts by scientists and nonscientists to show that religion and science were or were not incompatible assured this. See the discussion in Henri Guerlac, "Science and French National Strength," in Edward M. Earle, ed., *Modern France* (Princeton, N.J.: Princeton University Press, 1951), pp. 81–105.
39. Scientists like Willard Libby and Linus Pauling have been involved in a number of purely scientific controversies, but the hostility aroused by them has evidently been far less than the hostility that has grown out of the controversy about the effects of radioactive fallout. See the sketch of Libby in Theodore Berland, *The Scientific Life* (New York: Coward-McCann, 1962), pp. 41–44.
40. Cf. David C. McClelland, "On the Psychodynamics of Creative Physical Scientists," in Harold E. Gruber, Glenn Terrell, and Michael Wertheimer, eds., *Contemporary Approaches to Creative Thinking* (New York: Atherton Press, 1962), pp. 141–174. Modern science was born during the religious wars of the seventeenth and eighteenth centuries, and a common characteristic of the early scientists was their hostility toward the divisions and "enthusiasms" of the sects. See, e.g., Max Casper, *Kepler*, C. Doris Hellman, trans. and ed. (New York: Abelard-Schuman, 1959), pp. 77–85; 217; 335.
41. A generation later, Kepler discovered the true author of the preface, which had been attributed to Copernicus himself by many persons.
42. The preface is quoted in full in Arthur Koestler, *The Sleepwalkers* (London: Hutchinson, 1959), pp. 564–566.
43. A somewhat similar argument was developed during the evolution controversies. Some opponents of the theory argued that it was only an "hypothesis" which

would remain so, if stringent enough methodological principles were accepted. Men like Lord Kelvin rejected the theory because it could not be stated in rigorous deductive or mathematical terms, and others rejected it because it was not a "causal" theory in the way such theories were interpreted by the philosopher William Whewell. See Alvar Ellegård, "Darwin's Theory and Nineteenth Century Philosophies of Science," in Philip P. Wiener and Aaron Noland, eds., *Roots of Scientific Thought* (New York: Basic Books, 1957), pp. 537–568.

44. Of course, the extension since Mach of the positivist approach to all concepts has obviously fundamental implications.

45. See the quotations collected by Caplow and McGee, note 2 above. The denial of the existence of controversy by many scientists makes it a difficult topic to study. In this respect it is similar to secrecy in science. Scientists may deny the presence of controversies to strangers but not to themselves, but there is little reason to believe that this was true of those I interviewed. Some of the scientists questioned on this topic were candid and gave no positive responses about controversies even after much probing. They probably denied the existence of such disagreements to themselves as well as others.

46. George Wald, "Life and Light," *Scientific American,* 201 (October 1959), 92–100; reply by Arthur Galson, 202 (January 1960), pp. 12; 15; rejoinder by Wald, *ibid.*, p. 15. A scientist may reply to his opponents in popular or general journals like the *Scientific American* or *Science* because he feels their readers cannot adequately judge the material presented. When papers are published in specialized journals, on the other hand, the scientist may feel that sufficient control is exercised by the recognition or lack of it of the scientific community, making it unnecessary for him to enter into open debate with his opponents. However, it also seems likely that editors of specialized journals deliberately discourage expressions of controversy.

47. For another expression of the aesthetic nature of mathematics, see Jacques Hadamard, *The Psychology of Invention in the Mathematical Field* (New York: Dover, 1954), p. 127 *et passim.*

48. Some of the discussions have been reviewed in journals of popular science. See, for example, the review by P. W. Bridgman of Louis de Broglie, *Non-Linear Wave Mechanics: A Causal Interpretation,* in *Scientific American,* 203 (October 1960), 201–206; and the review by J. R. Newman of David Bohm, *Causality and Chance in Modern Physics,* in *Scientific American,* 198 (January 1958), 111–116. The "Copenhagen" interpretation of quantum mechanics appears also to be opposed in the Soviet Union, for reasons stemming from political ideology. See Gustav Wetter, "Ideology and Science in the Soviet Union: Recent Developments," *Daedalus,* 89 (1960), 581–603.

49. Cf. Thomas S. Kuhn, *The Structure of Scientific Revolutions, op. cit.*, p. 162.

50. Cf. Albert Einstein's by no means hostile remarks; he wrote that the scientist "must appear to the systematic epistemologist as a type of unscrupulous opportunist." In Paul A. Schilpp, ed., *Albert Einstein: Philosopher-Scientist* (Evanston, Ill.: Library of Living Philosophers, 1949), p. 684.

51. *The Structure of Scientific Revolutions, op. cit.*, ch. XI.

52. Cf. Thomas S. Kuhn, "The Caloric Theory of Adiabatic Compression," *Isis,* 49 (1958), 132–140.

53. Thomas S. Kuhn, *The Structure of Scientific Revolutions, op. cit.*, pp. 140–142. The definition offered by Boyle was essentially that elements were substances which could not be broken down into simpler substances by chemical analysis.

54. *Ibid.*, p. 138.
55. Significantly, no such instructions could be found in English or American journals; the Germans have a longer history of scientific quarrels.
56. Two American Nobel-Prize-winning experimental physicists, Robert A. Millikan and Arthur H. Compton, argued intensely for many years about the origin of cosmic rays. Millikan was allegedly the last man to admit that he was wrong. Significantly, he doesn't discuss the matter in his *Autobiography* (New York: Prentice-Hall, 1950).
57. Most scientists are unlike Kepler, who added a strongly critical preface and notes to the second edition of his *Mysterium Cosmographicum*, in which he made statements like, "this is not at all true," and "the reasoning of the whole chapter is wrong." See Koestler, *The Watershed, op. cit.*, pp. 62–64.
58. E. S. Pearson, *op. cit.*, 28 (1936), 229–233. Late in his life Louis Agassiz expressed some doubts about his earlier views on Darwinian evolution, although he still attempted to fuse his religious convictions with his convictions about the origin of species; see Lurie, *op. cit.*, pp. 373 f.
59. *The Positive Philosophy*, freely trans. and condensed by Harriet Martineau (London: John Chapman, 1853), 2, 152 f.
60. *The Origin of Species*, (6th ed.; New York: Mentor, 1958), p. 444.
61. Planck, *op. cit.*, pp. 33 f. Planck exaggerates a bit here; Ostwald and Mach eventually came to believe in atoms, partly as a result of Einstein's work on the quantum and Brownian movement. With respect to their opposition to atomic theory Einstein wrote, "This is an interesting example of the fact that even scholars of audacious spirit and fine instinct can be obstructed in the interpretation of facts by philosophical prejudices. The prejudice—which has by no means died out in the meantime—consists in the faith that facts by themselves can and should yield scientific knowledge without free conceptual construction." Einstein in Schilpp, *op. cit.*, p. 49. It is ironic that Einstein died unconvinced by modern quantum mechanics and that with his death much of the support for his position also died.
62. Cf. Kuhn, *The Structure of Scientific Revolutions, op. cit.*, pp. 89 f.
63. See, for example, the comments about opposition to relativity among physicists in 1913, by Edmund T. Whittaker, "Albert Einstein 1879–1955," in *Biographical Memoirs of Fellows of the Royal Society* (new series, London: Royal Society, 1955) I, pp. 36–67; and the comments by Emilio Segrè on the struggles in advancing the new physics he and young men gathered around Fermi had in the nineteen twenties, "Biographical Introduction," *Enrico Fermi: Collected Papers* (Chicago: University of Chicago Press, 1962), pp. xxviii–xxxiv. Other scientists have expressed themselves in more general terms about the conflict between youth and age in science; see Bernal, *op. cit.*, pp. 115, 391; and Samuel Goudsmit's comments in Daniel Lang, *From Hiroshima to the Moon* (New York: Simon and Schuster, 1959), pp. 234 f.
64. Compare Karl Mannheim's concept of generation-entelechies, although Mannheim himself argued that they would not occur in the natural sciences. *Essays on the Sociology of Knowledge* (London: Routledge and Kegan Paul, 1952), p. 319. Mannheim was led to this probably erroneous conclusion because he did not perceive that revolutions could occur in the natural sciences or that consensus might be difficult to achieve in them.
65. J. Bronowski, *Science and Human Values* (New York: Julian Messner, 1956), pp. 75, 87.

VII

THE FUTURE OF SCIENCE

This work has been concerned with what might be called the "classical" organization of science. Most of the scientists studied have been formally independent members of an autonomous community of colleagues. The thesis presented here is that the solidarity of this community and the conformity of its members is secured through intensive socialization and a complementary system of social control. Social control is exercised primarily through the reward of social recognition for contributions of information. This type of control reinforces the commitments of scientists to higher social norms, but it also makes the application of these norms flexible. As new goals and techniques are discovered, scientists are induced to accept them by their desires to receive recognition.

This view of the organization of science accounts for behavior laymen may find difficult to understand—behavior like that of the scientist who, not motivated by desire for wealth or power, spends years of work on a "small" point of no practical importance. It also leads to the discovery of the characteristic internal strains of science. The theory implies that social recognition is a scarce reward and that scientists will compete for it. Scientists compete for priority in making discoveries, and the major types of deviation to which this leads are secrecy and hasty publication.

When scientists are formally independent, then the short run co-ordination of the efforts of more than one of them, when it is felt to be necessary, must be arranged by the participating scientists themselves. Such co-ordination typically involves the collaboration of peers, an arrangement in which they agree to share recognition. In addition to this type of "partnership" arrangement, scientists form "apprenticeship" relations with students; both types satisfy needs for assistance without compromising the independence of scientists. The provision of professional assistance in formal organizations, i.e., "unfree" teams of collaborators, may threaten informal organization based on the exchange of information for recognition.

Perhaps the major type of disorganization in science as it is classically organized is anomy. Anomy exists when scientists are highly specialized but when their work is not interdependent; it means that scientists cannot make decisions that will be rewarded with recognition. Anomy is one of the major sources of scientific ritualism and retreatism. It is also a source of conflict within disciplines. The rebellion of a specialty against the discipline in which it is formally associated may also result from other causes. Rebellion involves goal conflict within disciplines, and it is often resolved by the structural differentiation of disciplines.

Cases of goal conflict, by providing an exception to usual patterns of behavior, illustrate how the formal organization of science is subordinated to informal organization. Usually the decisions of authorities in formal organizations follow from, and are consistent with, the informal organization of science as manifested in the award of recognition to individuals and prestige to specialties. When goal conflicts occur, this may not be true, and to many scientists the decisions of authorities will seem arbitrary and illegitimate. The same effect may be produced by another type of disorganization, that associated with substantive disputes. Differentiation is not a possible resolution to this kind of strain; the most common type of control is the voluntary disengagement of disputing parties, although in some cases the expression of controversies may be suppressed.

The conclusions reached here, on the basis of a small sample of scientists in a restricted area of science, probably apply to other branches of science as well: such field sciences as geology and nonexperimental zoology, chemistry, which was not treated in detail here, and the biomedical sciences. Attempts to disprove the theory advanced here probably would involve greater emphasis on the formal organizations in which scientists work and a closer examination of the relations between basic and applied research.

First, some critics may feel that I have exaggerated the importance of informal organization in science. Such critics might attempt to show, in contradiction to the evidence gathered for this study, that competition for priority is less important to practicing scientists than competition for position and material rewards and that the former competition derives its importance from the latter. They might also attempt to show that the struggle for power in university departments and other research organizations depends almost entirely on organizational factors and that goal conflict and substantive disputes, as they were described here, *follow* from such organizational conflict, serving, as it were, as "ideological masks" for them. Any approach emphasizing formal organization in this way would profit from a closer examination of nonuniversity organizations that conduct basic research.

Second, some critics might argue that the organization of basic research cannot be adequately treated apart from the organization of applied research. Major innovations in scientific thought may be made by applied scientists who are not part of basic research organizations; this was true in the medical sciences in the nineteenth century[1] and may be true in many scientific fields today. Perhaps future developments in scientific theory and the organization of science will be initiated by such occupational types as the physicist in industry, the economist in a school of business, and the psychologist in a mental hospital. Such applied scientists will be in positions where they may be strongly influenced by nonscientists. To the extent that nonscientists have this importance in the organization of science, the ideas I have presented will need to be qualified.

The need for these types of sociological research is more or less implied by the study reported here. The need for other types of sociological research has been either stated or clearly implied where appropriate in the text. In addition to sociological research, the views presented here imply the need for historical and psychological research. Historically, we need to know where and how science became professionalized and how the system in which information was exchanged for recognition became established. Psychologically, we need to know how, if at all, the occupational personality of the scientist is linked with the system of rewards through recognition in science.

These are some of the detailed research implications of the present study. In addition, the study has some implications for the future of science.

The Future of Science

The central problem facing society with regard to all of the professions is the same: how can they be controlled without having their effectiveness destroyed? Uncontrolled professions may not fulfill their responsibilities, but controlling professions may make it impossible for professionals to perform their tasks and may destroy their commitment to the values they are expected to uphold. The recurrent dilemma between freedom and control also applies to the position of science in society.

Most writers have emphasized one horn of the dilemma, the danger that science may be corrupted and may lose its autonomy in the near future. To some extent this may follow from its very success; the demonstrated potentialities of science, especially in weaponry and medicine, have vastly increased its public support. Scientists have received far higher material rewards in recent years than they used to, and it is alleged that they have been corrupted, oriented more to such rewards than to the recognition of their colleagues. In addition, science has become more dependent on such support. The threats to the independence of scientists posed by their dependence upon access to scarce and expensive facilities have been discussed in Chapter III. What were once individual decisions are now collective decisions often strongly influenced by the larger society. Finally, increased specialization makes it more difficult for scientists in different disciplines to act together. Science now consists of many more or less incommensurate disciplines. They do not present a unified world view to the larger society; they cannot agree on courses of instruction for university undergraduates; and they compete for the limited resources available to science.[2] *Between* disciplines the system of colleague control described in this work is largely ineffective. The result is almost necessarily the concentration of power in the hands of organizational leaders—university presidents and deans, foundation heads, and federal administrators.

Considerations such as these have led Norman W. Storer to write:

> I am suggesting that the new position of science in society has engendered internal conditions which are rapidly altering its entire structure. The two sources of change—increased support from outside and increasing growth inside—are operating to open wide the older, closed-system scientific community that we may be still assuming or hoping will be preserved. I suggest that it will *not* be preserved, and that we must accept this and bend our efforts toward preserving what we can of it in the new situation which is nearly upon us.[3]

This is one view of the future of science, a view heavily influenced by emphasis on applied science and on disciplines, especially nuclear physics, that are highly dependent upon public suport. But many of the factors leading to the corruption and heteronomy of science will, under other conditions, lead to the development of irresponsible science. Specialization, numerical growth, and increased public support may lead to the development of autonomous communities of specialists, well-insulated from the larger society, who transform their own limited goals into ultimate values and ignore their public responsibilities. One might emphasize such a view of science if he started with such fields as mathematics and astronomy. Such a view seems plausible when one reflects that artists and scientists are among the few groups in modern society who can advance their own goals as ultimate values; "art for art's sake" and "knowledge as an end in itself" are still advanced as slogans when other groups in society no longer dare express such similar views as "the public be damned."

These are two views of the future of science: on the one hand, a prediction that the spirit of science may decline while its forms continue to expand and, on the other hand, a prediction that science may become the province of specialized cults that will be irresponsible from the perspective of the larger community. The same views have been expressed about professions generally. Some writers have suggested that the professions will decline as their members are absorbed into formal organizations. Others have suggested that occupations in general will become professionalized, and one can extrapolate from this a view of society in which common interests are neglected in the face of the special interests of autonomous groups—a rigid society in which many guildlike groups exercise veto power over collective decisions.

Neither outcome is likely. The tension between professional autonomy and the control of professions leads to the development of organizational forms that can accommodate both. Industrial and governmental organizations have found forms capable of accommodating both their own interests and those of their scientific employees;[4] and American society, through forms such as the contract system, appears to be finding ways of accommodating scientific institutions to government.[5] Tensions continue to exist both within institutions and between them. Such tensions need not be a cause for deep concern: it is not tension but the absence of tension that is symptomatic of the loss of values.

NOTES

1. Joseph Ben-David, "Roles and Innovations in Medicine," *American Journal of Sociology*, 65 (1960), 557–568.
2. Actually, there has so far been relatively little interdisciplinary conflict over research funds. D. S. Greenberg has suggested three reasons for this. First, the amount of money available to science has been growing so rapidly that few disciplines are seriously deprived. Second, "In part this can probably be attributed to the fragmentation of federal support among numerous federal agencies, and the consequent lack of any battlefield where, for example, the biologists might have it out with the chemists. It is, in fact, far easier for subdivisions of a discipline to struggle against each other for the favor of the agency that provides the bulk of support for the overall discipline." Third, scientists usually seek to avoid administrative responsibilities and organizational conflict. "Money for Science: NAS Studies Likely to Have Large Influence on Future of Government Support," *Science*, 146 (1964), 508–509.
3. "The Coming Changes in American Science," *Science*, 142 (1963), 466. See also Alvin M. Weinberg, "Impact of Large-Scale Science on the United States," *Science*, 134 (1961), 161–164; Edward Speyer, "Scientists in the Bureaucratic Age," *Dissent*, 4 (1957), 402–413; and Derek J. de Solla Price, *Little Science, Big Science* (New York: Columbia University Press, 1963), ch. 4.
4. See William Kornhauser, *Scientists in Industry: Conflict and Accommodation* (Berkeley: University of California Press, 1962).
5. See Don K. Price, "The Scientific Establishment," *Science*, 136 (1962), 1099–1106.

INDEX